WORLD RELIGIONS AND CONTEMPORARY ISSUES

World religions and contemporary issues

HOW EVOLVING
VIEWS ON
● ECOLOGY,
● PEACE, AND
● WOMEN
ARE IMPACTING
FAITH TODAY

BRENNAN R. HILL, PH.D.

TWENTY
THIRD 23rd
PUBLICATIONS
NEW LONDON, CT 06320
WWW.23RDPUBLICATIONS.COM

Twenty-Third Publications
A Division of Bayard
One Montauk Avenue, Suite 200
New London, CT 06320
(860) 437-3012 or (800) 321-0411
www.23rdpublications.com

ISBN 978-1-58595-913-6
Library of Congress Catalog Card Number: 2013934664
Printed in the U.S.A.

DEDICATION

To my beloved wife of 40 years, Marie

TABLE OF CONTENTS

ACKNOWLEDGMENTS

I want to extend my sincere appreciation to those who assisted me with this book. Gratitude to my wife, Marie Hill, and to Sue Goldberg for their expert assistance in editing. Many thanks to Michael Strauss, Keith Egan, Rabbi Lewis Kamrass, Rabbi Abie Ingber, Peter Phan, Aman Choudary and David Loy for their invaluable scholarly assistance. Deep appreciation to the librarians at Xavier University and OhioLink for their dedicated service.

》 *Links to all the web pages referenced in this book can be found at*
http://store.pastoralplanning.com/woreandcoisi.html

An Introduction to Religion
and Its Connection with Social Issues

Jean Donovan wasn't known as a very "religious" person during her school days in Cleveland at Case Western Reserve University, where she received her master's in economics, or at the firm where she worked as an accountant. She was known as a headstrong and unorthodox young woman, who came to work on a Harley and liked to party with her boyfriend. Jean was a searcher. Her search took a new turn at an office Christmas gathering when she began to feel that something was wrong—somehow she wasn't where she should be. So she took a trip to Ireland to talk with an old mentor of hers who had been a missionary. The two talked through the night about the needs of the poor and about abandoned and starving children. Gradually, Jean came to realize that she was called to be a missionary, and after several years of training she went off to El Salvador. There, she was caught in a vicious war, and she discovered that to help the poor refugees fleeing the violence was to be branded as a revolutionary or a Communist. It became more dangerous for Jean as she rode from village to village on her motorcycle, guitar over her shoulder, to bring food, clothing, and some fun and entertainment to the young.

Gradually, Jean became a marked person. Her name was put on a death list. One night, Jean and several other missionaries were kidnapped by some soldiers, raped, murdered, and buried in a shallow grave. Jean had paid the ultimate price for her faith and for heeding the call to be a missionary to serve the poor and oppressed. She joined the ranks of martyrs for their religious faith, whether it be the Jews during the Holocaust, Christians during the persecutions in communist Russia and China, the Muslims in Bosnia, or the Buddhists of Tibet.

In this text, we are going to discuss those countless individuals who for thousands of years have carried out their search for the answers to ultimate questions within the religions of the world. By way of introduction, we are going to explore four questions: What is religion? What are the components of religions? Why study religion? Are the religions of the world concerned about social and political problems?

WHAT IS RELIGION?

The word "religion" has many meanings and has been applied to extremely diverse movements. Some have suggested that when discussing this area we might even need some other word that could be more accurately applied to diverse religions.[1] Following this approach, many Jewish scholars do not use "religion" to describe their tradition. The great Christian theologian Karl Barth maintained that Christianity is so unique that it should not be included among other religions.[2]

The word **religion** is derived from possibly three different Latin words, each of which offers some insight into the nature of religion. The first is *relegere*, which means to "constantly return to," or "to conscientiously observe," both of which can be seen in religious observance. The second Latin word is *religari*, which means "to be bound to," or "to be tied into," which in this context might refer to commitment to a religious tradition. And the third word is *re-eligere*, which means "to choose again," perhaps referring to religious conversion or returning to one's origins and goals. All of these meanings are useful in coming to understand the various dimensions of religion.[3]

Elusive to Define - There have been many attempts to define religion, but most fall short when it comes to including all the different traditions.

If religion is defined in terms of belief in a God, a problem arises when one realizes that even those religions that refer to God usually differ widely on what each means by God. Even in the so-called Abrahamic religions—Judaism, Christianity, and Islam—Adonai, the Trinity, and Allah have diverse meanings. Moreover, other religions such as Hinduism and Shinto speak of many gods and goddesses, each with a unique meaning. In Theravada Buddhism or Jainism, God does not come into the picture at all.

A similar situation arises if religion is defined in terms of a commitment to a set of beliefs. These beliefs can be so conflicting that members of one religion might consider other religions to be false and call members of those religions pagans, heathens, or infidels. Even within one religion, there can be such controversy about beliefs that some members are thrown out as schismatics, heretics, or false believers.

Many of the well-known definitions of religion seem to fall short. In the last century, the "father of Modern Protestant theology," Friedrich Schleiermacher (1768–1834), defined religion as "the feeling of absolute dependence."[4] Critics of this approach point out that religion includes much more than feeling, and some who value their freedom and independence balk at such a definition. Another classical theologian, Paul Tillich (1886–1965), defined religion as "ultimate concern for the ground of Being."[5] Here, critics answer that religion often entails much more than "concern," and others find it hard to relate personally with a "ground of Being." William James, the brilliant expert on religious experience, defined religion as "the feeling, acts and experiences of individual people in their solitude…as they stand in relation to whatever they may consider the divine."[6] While valuable, this definition doesn't seem to include the communal aspect of religion or the importance of **ritual**. Victor Frankl, a Holocaust survivor, said that religion was "a function of the spiritual unconscious, which is the source of the will to meaning."[7] Frankl's understanding was a major contribution to the psychology of religion, but it seems to neglect the conscious or mindful conduct toward belief as well as the activity of theology. All of this leads one modern scholar to conclude that "because of our increased awareness of the variety of religious experience found in the world, it is well nigh impossible to formulate a comprehensive theory of religion that will encompass this variety...."[8] In short, there is simply no one-size-fits-all definition or theory of religion.

Even though religion is difficult to define, most people have some notion and usually some experience of religion. Defining religion would seem less important than the ability to recognize it when we see it, and that usually means encountering some phenomenon that has something to do with the supernatural, the sacred.[9] And though religions might be quite diverse, they do have what the great philosopher Ludwig Wittgenstein called "family resemblances"—many parallels and similarities. Though religions differ, they still seem to be somehow related to the phenomenon of religion. This phenomenon can be looked at from different perspectives.

Perspectives on Religion - During the last several centuries a vast number of studies took a broad look at religion in the context of politics, liberation movements, economics, and international relations.[10] We will give a brief overview of some of the classic approaches to religion.

》 **See Huston Smith discuss religions as classical languages of the human spirit:** http://www.youtube.com/watch?v=EHE59oYEuKc

In the 19th century, many extensive studies on the history of religion appeared. Colonialism in different parts of the world produced extensive excavations, and new critical tools were developed for the analysis of sacred texts. Scholars studied the myths, totems, and beliefs of ancient religions, often attempting to discover the origin of religion itself. While these studies provided a great deal of interesting background to so-called "primitive" religions, none was really successful in discovering how religion originated. Moreover, many of the so-called "primitive" religions examined, such as the beliefs of the Aborigines of Australia, turned out to be rather sophisticated. Today, there are few works done on the history of religion as such, but there is much interest in studying the histories and developments of individual religions. This text will make much use of this historical research.

• The Sociology of Religion

Emile Durkheim was one of the pioneers in the sociology of religion. He took the position that religion was a product of the social experiences that early clans had around their totems, which for them were symbols of the sacred. Studying groups in Central Australia, he observed the great enthusiasm the natives experienced in their totem ceremonies and concluded that this marked the origin of religion. Durkheim's theories were seriously challenged, but nonetheless his work as well as other studies on myths (stories that express religious beliefs) and symbols of so-called "savage minds" were very influential on later scholars investigating religion in its social context. These scholars discovered that culture and religion deeply influence each other. Paul Tillich, a significant Lutheran theologian, has pointed out that it is culture that raises the profound questions that religion, in turn, attempts to answer.[11] The work of Max Weber also stands out. Weber is well-known for his study of Protestantism, its association with the development of the "work ethic," and how this work ethic is connected to the development of capitalism in Europe and the United States.[12] Today, the sociology of religion has become an advanced science that sheds light on the development of religions in many cultures throughout the world and offers insights on such issues as gender, morals, social mores, growing secularization, and patterns of religious practice. Modern sociology of religion points to many social issues, such as discrimination, inequality, and oppression, which need to be addressed by religion.

As we discuss the various world religions in this text, we will see how they are indeed addressing serious social issues such as the devastation of the places where people live, the violence and oppression to which many are subjected, and discrimination against women.

Some social critics have been hostile toward religion, saying that religions and God were mere projections, made to satisfy human needs for security and meaning. Karl Marx picked up on this and maintained that religion was "the opium of the people," a phenomenon that arises out of poverty and want. Religion moves people to accept their lot in life and blinds them from recognizing the real source of their suffering, the capitalistic elite who enslave them. This kind of thinking led revolutionaries to believe that if they were to arouse the people to revolution, religion would have to be eliminated. Thus came the arrests, imprisonment, and even execution of countless religious people under Communist leaders Vladimir Lenin and Joseph Stalin in Russia and Mao Tse-Tung in China.

• The Psychological Approach

Some scholars look at religion from the psychological point of view. Humanists such as Erich Fromm (1900–1980) view religion as just another expression of human capacities and have no problem with religion as long as it makes people better human beings.[13] Carl Jung (1875–1961) places the religious capacity within the unconscious, a capacity that is to be drawn toward religious experience. Others such as Victor Frankl (1905–1997) say that humans have a unique "spiritual unconscious" that drives them to seek ultimate meaning. To put this another way, humans have a unique capacity to connect with the transcendent or the supernatural, and religion is a result of the activation of this capacity. In his classic treatment of religion, Rudolf Otto (1869–1937) maintained that humans have a "creature consciousness," a "faculty of divination"[14] that enables them to experience the "holy," the "sacred," with a sense of awe, fascination, and attraction. For some psychologists, humans are religious by nature, possessing some natural or perhaps even "supernatural" capacity to search for, reach out to, and connect with transcendent reality.

>> *See* **Victor Frankl: Ultimate Meaning and Religion**
http://www.youtube.com/watch?v=OaQH-xACGK4&feature=related
and **Victor Frankl Documentary**
http://www.youtube.com/watch?v=YT7xA3YN57Q&feature=related

Victor Frankl was a Holocaust survivor and from his experience offered many valuable insights into religion, especially through his classic book *Man's Search for Meaning*.[15] He says that he learned in the horrors of Auschwitz that those most likely to survive were those who had meaning and purpose in their lives, things they wanted to accomplish. He quotes German philosopher Friedrich Nietzsche: "He who has a why to live can bear with almost any how."

Survivors of the concentration camps had healthy concerns outside themselves, and even though they couldn't change their horrible situation, they could change the way they perceived it. Even those who perished could enter the gas chambers upright, with the "Shema Yisrael," the treasured Jewish prayer, on their lips. Frankl was the only one in his family to survive the camps, and after his release he developed a therapy to help people gain meaning. He often worked with handicapped people and learned that they were happy and well-adjusted when they had a purpose in their lives. Frankl maintained that loss of meaning and purpose is often the cause of suicide. His formula for mental darkness was:

$$D = S - M.$$

$$Despair = Suffering - Meaning$$

Another Holocaust survivor seems to prove Frankl's point. Corrie ten Boom, a Dutch woman, survived the camps. She had promised her sister, who died in the camp, that she would travel the world as an evangelist preaching the power of love and forgiveness as opposed to the hatred that had caused all the destruction during the war.

There has been a strong alienation from religion in some psychological schools, especially those linked to Sigmund Freud (1856–1939). Freud maintained that religious practice is a symptom of an obsessive neurosis, a childish clinging to an "illusion" that there is a father who will protect us.[16] From this point of view, there is nothing about religion that a good therapist can't cure.

>> See **Freud on God**
http://www.youtube.com/watch?v=qHjhv6SOjC4&feature=related
and **The Question of God: Freud /Lewis**
http://www.youtube.com/watch?v=ymjuxVPBZYc

• **The Theological Approach**

Theologians of all faiths also have their theories about religion. Some approach the topic philosophically, others historically or from the point of view of anthropology. Theologian Karl Rahner (1904–1984) takes the latter approach. Rahner maintained that human beings were created as questioners. Humans are the only living creatures who ask questions: "Why is this so? Where does all this come from and where is it going? Why did this happen? Why am I here, and what am I supposed to be doing?"

>> See **Karen Kilby on Rahner**
http://www.youtube.com/watch?v=ZWeqGTaxJFM
and **It's a Graceful Life:**
http://www.youtube.com/watch?v=j82gDXRdzf8

Rahner believed that our questioning nature is always open-ended, that we are always looking for more answers. At the same time, he believed that there are limits to what we can learn, even with regard to natural phenomena. Rahner pointed out that human questioning can ultimately turn to God, both as the beginning and final end of our questioning. God, then, becomes the Final Answer, the Ultimate Truth. Yet, the answers we receive in this life are still limited, illusive, tentative, and partial.[17]

Rahner held that humans were created to be searchers, always seeking meaning, purpose, excitement, and pleasure—always looking for the truth. A refugee mother from war-torn Somalia searches for a place where she and her children can be safe and have shelter, food, and water. A little girl orphaned by the earthquake in Haiti searches for someone who will love and care for her. A young American joins the service and goes off to Iraq, hoping that he or she will mature and be able to face the tests of battle. Students go off to college, searching for understanding, new friends, a future career. Rahner taught that humans also have the God-given capacity to search for an Ultimate Reality, the Mystery that many call God. He uses the image of the horizon. The horizon is always with us, stretching in all directions. It is a backdrop for our movements on land, sea, and in the air. The horizon is always the beyond, wherever we go, always ahead of us, and even though we might not notice it, it is ever present. Rahner saw God as a horizon, a context and background for all of life—always with us and yet always ahead of us and unattainable. It is that Reality, against which we see everything else, to which we can reach out to.

David Tracy, an outstanding American theologian, took the title of his classic theological treatise from a poem by Wallace Stevens (1879–1955), which reads in part: "Oh! Blessed rage for order, pale Ramon, The maker's rage to order words of the sea, Words of the fragrant portals, dimly-starred, And of ourselves and of our origins..."

Tracy pointed out that belief in God can be the ground or horizon of our existence and enables us to live with apparent absurdity. He wrote: "That basic faith in the worth-while-ness of existence, in the final graciousness of our lives even in the midst of absurdity, may be named the religious dimension of our existence."[18]

>> **See David Tracy:**
http://www.youtube.com/watch?v=2_fu1bp25Y4

This human search or quest has been given various goals by scholars. Rudolf Otto (1869–1937) proposed that the quest is for the "holy," that mysterious dimension of reality that can be both attractive and at the same time frightening. Mircea Eliade (1907–1986), an outstanding expert on the history of religious ideas, maintained that the human (homo religiosus) search is for the sacred—that religion arises out of the experience of the sacred time and sacred space through myths and rituals, as opposed to the daily human experience of the profane, which is dissatisfying and without value.

>> **See Joseph Campbell: Myth:**
http://www.youtube.com/watch?v=VgOUxICCHoA

THE COMPONENTS OF RELIGION

Often scholars have dealt with religion only from "the outside." Research tends to be scientific and objective, and that approach certainly has its value. Another way to approach religion is from "the inside," looking at the components of religions, listening to the concerns, and noticing the observances as well as the actions of those committed to these religions. These components need not apply to all religions in the same way, but they are part of the "family resemblances" among religions. These components include the beginnings and inspired founders, communities and divisions, sacred writings, beliefs, prayers, symbols, rituals, practices, and actions of the members. These are at the heart of this mysterious phenomenon called religion.

Beginnings and Founders - All religions have a beginning. Sometimes that beginning is lost in the mists of time, as is the case with Hinduism and Shinto. Often religions have a specific founder, an inspired individual with a unique message that moves people to embrace their teaching and live accordingly. That person is clearly identified in Buddhism (the Buddha), Christianity (Jesus), Islam (Muhammad), and Judaism (Abraham and Moses). At times this person stands among other gurus who have contributed to the tradition (Mahavira in Jainism and Nanak in Sikhism).

Religions seem to begin with divine revelation to an individual or individuals. These inspired revelations are usually considered to be authentic and direct communications from God and serve as the basis for the formation of a new religious movement. Often the revelation to these individuals encourages them to challenge the religious beliefs of their time. Buddha's enlightenment moved his followers to challenge the established beliefs of the Vedic tradition, later known as Hinduism. Both Abraham and Moses moved their people to believe in one God, as opposed to the polytheism of their time. Jesus seems to have wanted to inspire the Jewish people to follow the best of beliefs and practices. And Muhammad was convinced that his final revelation served as a correction to the errors that had developed in both the Jewish and Christian traditions. Jainism challenged the tenets of Buddhism, and the Sikhs claimed to go beyond both Hinduism and Islam.

Religions usually have individuals who provide leadership and even continue to have their own revelations. Jewish prophets continued to receive revelations from God, warned their people when they were wandering off course, and pointed out the evils and injustices outside and inside their communities. Religions like those in the Americas had shamans who had access to the spiritual world and could thereby guide their people. Other religions such as Hinduism, Christianity, Shinto, Taoism, and ancient Israel had priests who led rituals and offered sacrifice. Some religions, such as Buddhism, Christianity, and Jainism, had monks and nuns who practiced their tradition in exemplary fashion. Many of these rites and practices still exist today.

Communities - Ancient religions often began in tribes or peoples who experienced contact with some divine figures. Religions such as Islam, Judaism, Hinduism, Shinto, and those of Native Americans arose out of tribal groups and even today are often identified with national cultures. Religions also began with religious leaders calling together followers, disciples who would establish a way of life according to the leader's teaching. Buddha made the sangha (community) an essential element of his tradition. Abraham and Moses established "the people of God." Muhammad gathered followers who were devoted to his revelations from Allah. Humans are social by nature, and religions therefore usually value the communal framework for the sharing of beliefs, rituals, support, and lifestyle.

Many religions propose a way of life for the community. For Christians, Jesus is the Way—and the Way was the earliest name for the movement; for Taoists, the Tao is the way of the universe, and disciples are to align themselves with this pattern; for Jews, the Torah is the way to sustaining the covenant with God; and for the Hindu, Karma (the eternal law) is the way toward Nirvana.

These religious communities evolved, had their disagreements, and usually divided into various factions: Theravada and Mahayana Buddhism; Sunni and Shia in Islam; East and West Christianity; Protestant and Catholic Christianity; Reform, Orthodox, and Conservative Judaism; White-clad and Sky-clad Jainism. These divisions usually were a result of disputes about the texts or interpretations of beliefs.

Sacred Writings - Some ancient religions, such as those in the Americas, passed traditions on in oral form in stories and songs that were never recorded. This was the case with so many of the religions of Africa, Australia, and the Americas. After the development of writing, religions gathered their oral stories and teachings, put them into script, and collected them into books of sacred writing. The Jews assembled the Torah; the Christians, the New Testament; the Muslims, the Koran and Hadith; the Buddhists, the Pali Canon and many collections of Sutras; the Hindus, collections such as the Vedas, Upanishads, and Bhagavad Gita; the Sikhs, the Guru Granth Sahib. These collections represent the inspired teachings, myths, songs, prayers, and stories of these communities.

Generally, these sacred texts are viewed as somehow divine revelations and inspired traditions, which usually began as oral teachings. These texts capture the visions of religions, their code of laws, often the procedure for their rituals, and the systems of beliefs. For those with a literalist bent, these texts are to be taken as codified and defined for all time. For those that are non-literal, these texts are received more as literature, shaped by cultures, always open to new interpretations and ever evolving.

Faith and Beliefs - It might be useful here to distinguish between faith and belief. In its broadest sense, **faith** means a commitment in trust. One might have a human faith in one's doctor or financial planner, or a religious faith in God. Faith is sometimes used to refer to a religious tradition, as in the "Jewish faith." It can also mean the acceptance of certain beliefs—a faith in life after death.

On a deeper religious level, faith is more relational. Paul the apostle wrote: "I have faith in the Son of God who loved me and gave himself for me" (Galatians 2:20). For Paul, this kind of faith implied a deep trust and commitment to Jesus and his teaching, a mission to spread this teaching, and even a willingness to die in order to be true to it.

Belief in its earlier etymology referred to an "act of loving" or "holding to be dear." Gradually, both belief and faith came to be identified with formulations, statements, and doctrines that are held by each religious community to be true, e.g., Christian beliefs and the Hindu faith. Today, there are serious efforts to see faith as relational and as a special commitment and to distinguish faith from beliefs, which may be defined as doctrines, teachings, and formulations.

There is generally a hierarchy to these beliefs; some are considered **dogma** or absolute, others negotiable. For instance, all Buddhists believe that the Buddha reached enlightenment during his lifetime. That belief is non-negotiable dogma for Buddhists. But Buddhists debate whether or not women can be ordained as nuns and whether anyone can become a Buddha. All religions have some beliefs regarding God, the virtuous life, and life after death. The religions of the East often believe that the High God is impersonal Spirit who can exist in all and all in God, while the religions of the West often believe in a personal God with whom people can establish a relationship or covenant.

All religions are concerned with some Ultimate Mystery. Mystery here does not mean some problem to be solved, but rather a Reality that is beyond comprehension or description. The words "immanent" and "transcendent" are often used with regard to God. **Immanent** usually refers to God being within all of reality. **Transcendent** generally means that God is beyond all reality. Some distinguish between the "inside God" and the "outside God." Aside from Theravada Buddhism and Jainism, religions describe this Reality as God (and Goddess). The Hebrew tradition insists that this Reality is One and refers to God as Yahweh, the nameless one. The Hindus call this ultimate God Brahma, and generally see other gods and goddesses as manifestations of Brahma. Some of the Hindu deities have **avatars** or incarnations; Krishna is considered to be an avatar of Vishnu. Muslims speak of Allah; some Native Americans speak of the

Great Spirit. Christians believe in one God, manifested in three persons: Father, Son, and Spirit. Jesus is seen as an incarnation of the Son of God. Each of these religions are quite diverse in the way they understand and relate to God.

Many religions believe in the spirit world. Animists—followers of animistic religions—believe that there is a life principle or spirit in non-human entities. Shinto reveres the "kami," spirits who dwell in sacred mountains, forests, and elsewhere. The Native Americans believe that spirits dwell in nature, in rivers, mountains, and animals. Their main God is often called the Great Spirit. Christians believe in the spirits of the dead and risen, in "the communion of saints," and in the inspiration from the Spirit, as well as the "fruits" and "gifts" of the Holy Spirit. In the Judeo-Christian tradition the word "spirit" is related to "breath," which is the sign of life. In these traditions, life is seen as "breathed forth" by God.

• A Virtuous Life

Commonly, religions see the way or path leading to a virtuous life that follows the laws of God. The Vedic traditions of India refer to the Sanatana Dharma, the eternal law, and submit to Brahma, whose law calls all to abandon evil and embrace good. For Buddha, the dharma is summed up in "do good and avoid evil" or "do no harm." This is spelled out in the Four Noble Truths (eliminating desires ends suffering) and the Eightfold Path (goodness in all aspects of life). The goal is to extinguish the selfish self, become the "no self," end all suffering, and reach Nirvana, which is the putting out of the fires of desire and the end of the notion of death. For Jews, the law is the Torah, with its 613 commandments and scribal laws for living. For Christians, the law lies in the Commandments and in the Beatitudes of Jesus. In the Beatitudes, two commandments call for love of God and neighbor, which arose from the Jewish tradition. Islam teaches the Five Pillars of the holy life—worship, prayer, alms, fasting, and pilgrimage.

Most religions call for such virtues as detachment, compassion, love, justice, and nonviolence. Most see the connectedness of all things and call for a reverence for life. Many imitate a benevolent God, a God who sustains and comforts and calls for worship and praise of this Ultimate Reality.

• *Life after Death*

All religions wrestle with the ultimate question surrounding death and afterlife. Early Judaism spoke of Sheol, a kind of half-life after death, as one awaits the final day of judgment. The ancient Pharisees, who served the common people as models of Jewish life, believed in the ultimate resurrection of the dead, whereas the Sadducees, wealthy nobles and often high priests, did not. Many Jews today are satisfied to live on in memory, in their accomplishments, and in their children. Often, Christians believe in resurrection to eternal life in heaven for good people and eternal punishment in hell for the evil ones. Karma is a common term in the Indian religions and usually refers to actions that determine one's rebirth. In Hinduism one can be subject to many rebirths until one achieves liberation (moksha) and moves on to Nirvana. The soul (atman) is the deathless part of the person and can be united with the Creator. In Buddhism, Nirvana is the goal, where one is free from the burning attachments of desires, free of suffering, and free of the concept of death. The Buddhist tradition does speak of some rebirths and describes levels of torture and hell. For Jains, all karma can be ruined by bad actions and exists as a sticky substance that holds the soul back. The ultimate goal is to cast off all karma and achieve a state of bliss. The Native Americans generally believed in an afterlife, but interpreted that differently, ranging from a belief in reincarnation to a belief in life in the sky or in the underworld.

>> *See* **Carl Jung on Death:**
http://www.youtube.com/watch?v=L0xlZm2AU4o

Prayer, Symbols, and Rituals - Religions commonly place value on some sort of communication with the Ultimate. One of the pillars of Islam is to pray five times a day, to praise Allah and acknowledge submission to Allah's will. Jews praise God in temple and synagogue, at table, and individually. Hindus practice many forms of yoga to center the energies of the body and find the divine within. Buddhists place value on meditation, not as a means of communicating with God, but as a means of purifying the consciousness from toxic thoughts and desires. The goal here is personal enlightenment. Christians have always stressed prayer as praise of God, petition, and request for forgiveness. Meditation, or reflecting on holy thoughts, as well as contemplation—entering into union with God—are also valued by Christians.

All religions have their **rituals** or actions used to express and celebrate their beliefs, including the seasonal dances of the Native Americans, the commemoration of Passover in Jewish homes, the celebration of Mass by Catholics, the Protestant attendance at service, and the chanting of the scriptures by the Sikhs. These devotions are designed to praise God, as well as transform those who celebrate them into virtuous people dedicated to following the path or way of their respective religions.

Religions have many sacred **symbols**, signs that point to the Ultimate and assist the devout to be in touch with their deities and mystery. Christians have sacraments, the sacred rituals of the church, to put them in touch with the power of Jesus Christ. Jews use symbolic foods such as unleavened bread and bitters to relive their liberation from Egypt during the first Passover. Totems have been widely used by religions, and their carvings bind people to their kin as well as to the images that are engraved on them. Buddhist monks and nuns wash and shave their heads to indicate the impermanence of all things and their detachment. They cherish the lotus flower, which rises in beauty from the mud, symbolizing the Buddhist search for purification.

WHY STUDY RELIGIONS?

Since the infamous attack on the World Trade Center on September 11, 2001, people have become more aware of the role that religion (in this case, extreme Islamic beliefs) can play in politics and international relations. Young Muslim extremists were willing to give up their lives and destroy the lives of thousands of innocent people in order to make a statement against capitalism and American influence in the Arab world. The United States' involvement in Iraq and Afghanistan that followed revealed further links between religion and violence, economics, and women's rights. We witnessed Sunnis killing Shiites, and Muslims who were both friendly and hostile toward American soldiers. Many innocent people were killed in these wars, and many young soldiers from the United States and their allies were brought home dead or broken in this struggle, with its political, economic, and religious issues.

Many Americans continued to feel more confusion or even alienation toward the Muslim world, often identifying it with terrorism. Suddenly in the winter of 2011, a new image of Muslims grabbed international attention. Hundreds of thousands of Egyptians gathered in Liberation Square in Cairo, Egypt, calling out for freedom and the resignation of their president, who had been in power for 30 years. Muslims from all walks of life, many of them young college students, professionals, even parents with their children, demonstrated nonviolently for eighteen days. Muslims prayed on their knees in the streets and joined hands with Christians—united in their protest for freedom. More than three hundred of the protesters were killed, but the numbers of protesters grew until they finally got their wishes—the resignation of the president and hopes for a new future for Egypt. These types of demonstrations spread to Tunisia, Yemen, Syria, Libya, Jordan, and other Arab countries. The world was left with new questions about the future of Egypt and, indeed, the whole Muslim world. New questions arose about the links between religion and world events, and hopefully new motivation for the better understanding of religions and their impact on the modern world.

How Aware Are We? A Personal Journey

Even though we live in an age of information and have instant access to international news and facts on any topic through the Internet, most people have a limited understanding of world religions. My wife and I have traveled the world and find this is universally true. In an international conference on Buddhism in Vietnam, we discovered that few of the Buddhist participants knew much about any religion other than their own. As we visited Muslims in universities in Turkey, India, Egypt, and Jordan, we found that both professors and students knew very little about other religions. We asked one professor in Cairo what he thought would happen if some Buddhist monks came walking down the street, and he said they would probably be arrested! We have been in India several times and have found that the Hindus we have interacted with have little understanding of any religion other than their own. Ask some of your fellow students what they know about other religions and you might have the same experience. We live in Cincinnati, Ohio, within a thirty-minute drive from a Hindu temple, a number of Jewish synagogues, a Buddhist temple, a Baha'i chapel, a Muslim mosque, and many Catholic and Protestant churches, yet many people here are not aware that so many diverse faith communities exist. Nor is there much communication among these congregations or communities, even those of the same religion.

Most people live in their own cultural and religious "bubbles," quite unaware of the religions around them. This can lead to a narrowness regarding other religions and even to some disturbing stereotypes. Some Catholic priests have sexually abused children. As a result of this and the many cover-ups by church authorities, some individuals have become wary of all priests. Quakers report that they are wrongly depicted as people who sit around and "shake" during their meetings, or even as individuals whose main diet consists of Quaker Oats. Native Americans are misrepresented as dancing around a campfire and can be thought to be high on peyote or alcohol. The Amish are often incorrectly portrayed as old-fashioned and outdated people who ride in horse-drawn buggies. Mormons are caricatured as men surrounded by a harem of wives.

Such stereotypes are simplistic and unfair, and cause people to lose respect for other religions. Even worse, such attitudes can foster deep prejudices and even hatred toward these religions. Such attitudes can divide, alienate, and even lead to violence. Adolf Hitler and many others in Europe held a prejudice and hatred for Jews. This fierce hatred led to the Final Solution, a Nazi decision to kill all the Jews in Europe. Millions of innocent Jewish men, women, and children were transported to concentration camps, where they were gassed and their bodies incinerated in ovens.

The study of religion, where one becomes more familiar with the lives and practices of other religions and engages in respectful dialogue with people of other faiths, can help correct such stereotypes and prevent dangerous prejudices. Such study can help people be more comfortable with those from other cultures and religions. It can open students to new friendships and, hopefully, can prevent the alienation that often leads to violence.

Preparing for a Global Society - Society seems to become more global each year. Travel has advanced to the point that within a few hours one can travel coast to coast and within a day be in any other part of the world. Television, cell phones, the Internet, and social media have connected people from all over the world. In Egypt, the stirrings of a revolution gained momentum through social media, and the news media allowed the world to follow its path. Through technology, the whole world turned in sympathy toward Japan when its people suffered an earthquake, tsunami, and nuclear danger all at once. The linked nature of world economies became obvious when the near collapse of American investment and banking firms had serious repercussions worldwide. The global nature of trade can be seen when we look at the labels on our clothes and on many of the other items we buy. Immigration is going on throughout the globe, to the point where European countries and the United States find that they need to better accommodate the many languages, religions, and cultures found within their borders. The growing Latino community in the United States has led many to learn Spanish, and China's prominence has drawn many to learn the Mandarin language.

The study of world religions can help students be better prepared to participate in this global society. Such study can help toward better understanding and communication with other students and with neighbors, as well as prepare them for the future workplace. Our universities and workplaces are increasingly becoming more diverse. Becoming better informed about world religions, as well as about the cultures where they are practiced, can help students be more successful in their education and in their future workplaces. Such study can be useful to employees who are more often called to travel to other countries or to conferences with people with many different backgrounds. Knowing something about the religious backgrounds of others can assist in understanding their values and worldviews.

Help with the Personal Search for Truth - As mentioned earlier, human beings are searchers for meaning and purpose. Religions have always been part of this human search. The study of religions puts students in touch with people who for thousands of years and in many diverse ways have tried to understand ultimate Truth and who have sought to live good lives. Such study offers an opportunity to better understand one's own religion, as well as provide new and interesting religious perspectives. Such efforts might suggest looking into religious communities that can provide support, mentoring, and opportunities for service. Those who are indifferent or even hostile to religion can find new insights into religion and, if nothing else, come to better understand the religious perspectives of their friends, classmates, and fellow workers.

ARE RELIGIONS CONCERNED ABOUT SOCIAL ISSUES, SUCH AS ECOLOGY, PEACE, AND THE WOMEN'S MOVEMENT?

Religions are often separated from the world. Some Buddhists, Jains, and even Christians are world-denying and advocate leaving the world to live in forests or monasteries, detached from worldly desires and concerns. For some, religion is a private matter, concerned with personal beliefs and salvation, and should not be involved with "worldly" issues such as hunger, poverty, or ecology. From this perspective, religion should be concerned about preparing for the afterlife, rather than focusing on this life.

At other times, it is the world that sets aside religion. The First Amendment of the U.S. Constitution states that "Congress shall make no law respecting an establishment of religion, or prohibiting the free exercise thereof." Ironically, this amendment, which was written to preserve religious freedom, has been used to separate religion from life. Examples of such separation include forbidding public prayer in schools and the outlawing of public displays on religious feasts such as Christmas, Hanukkah, or Ramadan.

Other movements struggle to separate religion from society. **Secularists** often view religion as irrelevant to society and want it set aside and to be silent about political or social matters. Such secularism is spreading quickly throughout Europe and is gaining influence in the United States. **Post-modernists** generally see religion as just another culturally produced reality that unrealistically holds to absolutes. Post-modernism often advocates that religions be set aside as antiquated and irrelevant.

Many of those committed to religion object to such separation of religion from the world and want to have a voice in political and social issues. They believe that religious freedom not only gives them the right to believe as they choose, but also a right to apply their values to everyday issues. For instance, the Religious Right has become a powerful force in American politics.

Edward Schillebeeckx, an influential Dutch, Catholic theologian, offered a strikingly new version of the Catholic expression "outside the church there is no salvation" when he declared "there is no salvation outside the world." It was his position that every religious person works out his or her salvation in this world, where God's saving presence is experienced and where people are called to serve one another. Many followers of religion believe that their adherents should be concerned about the self and also the neighbor. Therefore, they profess religious beliefs and values that call for action on behalf of the needy and oppressed. Many follow the Golden Rule, which in one form or another exists in a number of religions. This "rule" obliges people to treat others as they would want to be treated, and therefore calls on religious people to be concerned about injustice, violence, prejudice, and abuse.

The modern era has seen numerous religious heroes that have been active in political and social issues. The Hindu Gandhi led his Indian people in South Africa and later in India in a strong and nonviolent resistance to colonial oppression. He organized marches, fasted, and led prayers and religious discussions until he liberated his people from British colonial rule. The Catholic nun Mother Teresa offered her life serving the poor, the homeless, abandoned, and dying in India. The Baptist minister Martin Luther King Jr. literally gave his life to gain civil rights for his people in the United States. The Tibetan Buddhist Dalai Lama has spent most of his life promoting peace in the world and struggling to gain freedom from Chinese rule for his Tibetan people. Throughout the text you will meet many other heroic activists from many religions who are committed to the struggle on behalf of the environment, peace, and equality of women.

The religions of the world are coming to realize that they represent millions, and in the case of Christians and Muslims more than a billion people each, and that such "people power" can make a significantly positive impact on the world. As we shall see throughout this text, religious traditions carry ancient beliefs about the sacredness of the earth and human people that could transform the world.

World organizations such as the Council for a Parliament of the World's Religions and the World Council of Religious Leaders address many social issues such as immigration, human trafficking, poverty, oppression of indigenous people, and many other issues. The Charter for Compassion has brought many of the world religions together to extend service to the needy of the world.

World religions are actively concerned about social issues today: prejudice, poverty, slavery, the sex trade, drugs, food and water shortages, oppressive prison systems, and many other issues. In this text, we are going to focus on just three of these issues: ecology, world peace, and the women's movement.

Religion and Ecology – A significant series of conferences on religion and ecology were organized by Mary Evelyn Tucker and John Grim at Harvard University from 1996–1998. Individual conferences were held relating each major religion to ecology, and additional conferences were held on environmental ethics and on discovering common ground among the religions. The conferences revealed that many different religions were growing in their concern about the environment, connecting their beliefs with these concerns, and producing activists to deal with the environmental crisis. A series of volumes was published from these conferences, and the whole endeavor gave impetus to colleges and universities to develop courses, programs, and degrees on the relation between religion and ecology.

In 1995 Prince Philip of England organized the highly influential Alliance of Religions and Conservation and offered Windsor Castle for their conferences. Leaders from all the major religions have gathered over the years to develop environmental programs based on the values of their religions. Most religions now have seven-year plans for programs on sustainability and are vigorously addressing the issues of global warming, climate change, and ways to preserve the natural environment through faith-based programs.

Religion and Peace - While it is true that the religions of the world have a history of warring over political and religious differences, most religions today strongly oppose violence and are vigorous advocates for peace. The notion of *ahimsa* or "do no harm" has always been strong in Hinduism, Buddhism, and Jainism. The Hebrew prophets often called for peace among the chosen people. Jesus advocated loving one's enemies and turning the other cheek when hit, and he told his disciples to put away their swords when they tried to protect him. Muhammad advocated peace among the Arab tribes and allowed violence only in self-defense. The Protestant Reformation spawned the Quakers, Amish, and Mennonites, all of whom have been strong advocates for pacifism. In nineteenth-century Russia, the great novelist Leo Tolstoy (1828–1910) was an influential promoter of nonviolence. His works profoundly influenced Gandhi, a Hindu leader of India, who was a strong advocate of nonviolence. Gandhi used nonviolence as a tactic in liberating India from British colonial rule. Martin Luther King Jr., in turn, was a follower of Gandhi and effectively used nonviolent methods in the struggle for civil rights for African Americans in the United States. Buddhists Thich Nhat Hanh (Vietnam), the Dalai Lama (Tibet), and Aung San Suu Kyi (Myanmar, formerly Burma) have all been heroic advocates for peace in our time. In South Africa, Christian archbishops Desmond Tutu and Denis Hurley and former South African president Nelson Mandela were harsh critics of apartheid and advocates for nonviolent resistance. As we shall see, all religions are today engaged in the crucial work of peacemaking. Major efforts are being made to stop the many wars that threaten the lives of so many around the globe. In this text, we will look at numerous activists in religions who are engaged, some risking their lives, in the struggle for nonviolence and peace.[20]

Religion and the Women's Movement - Throughout history there has been a recurring cycle where women, usually wealthy and educated women, have at times been able to enter areas of leadership and prominence. Those gains were usually short-lived and patriarchy was reestablished as the norm. In ancient Sumer, the earliest known civilization in the world, located in present-day Iraq, women gained considerable equality, only to relinquish it after conquests by Babylon and Assyria. Women in ancient Egypt gained many rights, only to lose these after the conquests by Greece and Rome. In ancient Greece women rose to high status, but lost their prominence during later periods. The philosophers Plato and Aristotle (ca. 428–322 BCE) both made statements about the inferiority of women that not only influenced Greek culture but also many later cultures in Europe.[21]

The domination and victimhood of women by men have been described in the writings of Virginia Woolf (1882–1941), which influenced the modern women's movement. This movement was also given impetus by American women such as Elizabeth Cady Stanton and Susan B. Anthony, as well as women in Britain, who vigorously campaigned for women's rights, especially the right to vote. The work of Simone de Beauvoir (1908–1986), a French writer, and the writing of Betty Friedan (1921–2006) in the United States, with her work *The Feminine Mystique* (1963), were also notable. As the movement evolved, women worldwide realized that their cultural situations and issues differed from the white, middle-class *feminist*. Black women began to discuss their issues as *Womanists*, and Latino women began to talk about their own issues as *Mujeristas*.

At present, the women's movements are widely diverse and sometimes divided. In the United States, much equality has been gained in education, business, the professions, the military, and politics. Current studies reveal that many women still do not receive pay equal to their male counterparts. Many also assert that they are regularly stopped by a "glass ceiling" that prevents them from achieving the highest position within their area of endeavor. Internationally, women of color, along with their children, are often predominant among the many refugees fleeing the violence in their areas. Commonly, women are subject to rape, oppression, and even enslavement in the sex trade. The impoverishment of women as well as sex abuse and domestic violence are among the major social problems worldwide.

Most religions in their original outlooks have viewed women as created by God with a unique dignity and as worthy of respect. The Hindu religion has its goddesses, who reflect the various powers of Brahma. Though these goddesses are usually consorts of the male gods and are largely interpreted by males, they provide Hindus with a basis for holding women in high esteem. Buddha, though he did not see women as equal to men, was able to be persuaded to have nuns ordained to the monastic life. In one of the Hebrew creation stories, both male and female are created in the image and likeness of God. Israel honors outstanding women like Sarah and Rebecca as the mothers of Israel. Miriam, the sister of Moses, is one of the leaders of the Exodus and is described as a prophet. Deborah is both prophet and judge and is called to bring God's word and justice to her people, and women have ruled as queens of Israel. By the time of Jesus, patriarchal social structures once again prevailed. Women had few rights, were not permitted to study Torah nor be priests, and were segregated in the synagogue and Temple. Much of this discrimination in Judaism and in other religions was due to female menstruation and giving birth, events that were perceived to render women impure. Jesus' reform called for a restoration of the dignity of women. Jesus radically broke with the dominant patriarchal structures by teaching women, choosing them to be his disciples, acknowledging their accompaniment of him to Calvary, and by calling them to be the first witnesses to his resurrection. Catholicism holds Jesus' mother Mary as well as Mary Magdalene in high esteem.

Muhammad's revelation set out to restore the dignity of women in the Arab world. He honored women, allowed them to pray with men, gave them more legal rights, consulted them, and forbade the killing of girl babies. Muhammad saw women as equal but different. Theirs was the private realm of the home and family. The realm of the male was to deal with the public matters of leadership, business, and defense. Jainism began with an inspiring view of women's equality and the role women can play in liberation. The Sikh vision of women gave them complete equality in all aspects of life. Taoism taught that the spiritual essence transcended sexuality and that there are females and males traveling in this lifetime, existing as the perfect yin and yang of reality. Even Confucianism, which in the past was often described as patriarchal, is now seen by some scholars as having been distorted by later practices. In truth, Confucianism seems to have much in it that can gain equality for women.

There seems to be a pattern in the history of religious traditions. Most religions in their beginnings have an enlightened, even radical, vision of women's equality. Eventually, patriarchal cultural structures once again prevail. Culture seems to trump religion, as cultural beliefs in women's impurity from menstruation, inherent weakness of mind and emotion, and their powers of seduction once again prevail.

Many significant advances have been made in world religions with regard to the dignity and equality of women. In Mahayana Buddhism many more nuns are being ordained and are gaining more rights to education. Protestant churches, for the most part, are ordaining women as ministers and bishops. The Reform Jewish movement has made great strides in recognizing women by ordaining women as rabbis and establishing equal roles in religious services and administrative positions. Within Islam as well as in Roman Catholicism, there are lively debates over the dignity of women and the roles they should be allowed to play in religion. As we shall see, the religions of the world include many women's movements that are working to achieve the dignity and equality that all women deserve in both society and religion.

SUMMARY

The term "religion" is complex, and it has always been difficult to establish a definition that applies to all religions. There have been many efforts to define religion, but usually these efforts fall short. Most definitions seem to include the human search for ultimacy. Religion has been studied from a number of perspectives, including historical, sociological, psychological, and theological. Hostile approaches to religion have described it as obsessive, as illusionary, or as a drug that causes passivity. There are a number of components of religion, including beginnings and founders, communities, sacred writings, beliefs, and rituals. The study of religion can be valuable for a number of reasons. Such study can offer students a broader worldview, help prepare them to live and work in a global society, and assist them in their personal search for meaning.

Religions have always been a way of life, a motivator for action. Today, religions are becoming more involved in world issues and are attempting to reassess their beliefs and values so that they can be active players in global concerns, especially in the areas of ecology, peace, and women's issues. Religions have the vision and values whereby the multitudes of people can be what Gandhi called the "soul force" to help make this a better world.

CHAPTER 1 VOCABULARY

avatar - The manifestation of a god, at times in human form

belief - Trust and assent to doctrine

faith - Trust and allegiance to a religious tradition

holy - Set apart, divine, consecrated

immanent - The presence of the divine within the world

karma - The eternal law

mystery - A reality beyond comprehension

myth - Story symbolizing religious truth

prayer - Communication with God

sacrament - A symbol of spiritual grace

sacred - That which is holy or consecrated

sacrifice - An offering to God to appease or petition

spirit - An invisible force or energy

supernatural - Beyond the natural laws or experiences

1. What do you think people might mean when they say that they are spiritual but not religious?

2. What is the difference between beliefs and religion?

3. Do you think that religion is intrinsic to human nature or extrinsic, e.g., cultural, family influences?

4. How would you respond to Marx's remark that religion is a drug or Freud's view that religion is based on illusion?

5. How would you explain the fact that some religions have a history of violence or that they have such high ideals, yet include histories of destruction and violence?

6. If religion is often considered to be a private matter, why is it that there are so many enthusiastic religious communities?

7. Do you think religious people should be involved in ecology, peacemaking, or other social issues? If so, for what reasons?

1. Do you think religious people should be involved in social issues at all? If so, for what reasons?

2. What do you think is the most urgent issue that Christianity should be concerned about? What Christian values would be applicable here?

SUGGESTED READINGS:

Corbett-Hemeyer, Julia. Religion in America. *Upper Saddle River, New Jersey: Prentice Hall, 2010.*

Crawford, Robert. What is Religion? *New York: Routledge, 2002.*

Deacy, Christopher and Elizabeth Arweck. Exploring Religion and the Sacred in a Media Age. *Burlington, Vermont: Ashgate, 2009.*

Haught, John F. What is Religion? *New York: Paulist Press, 1990.*

Hinnells, John R., ed. The Routledge Companion to the Study of Religion. *New York: Routledge, 2010.*

Klassen, Pamela et al. Women and Religion. *New York: Routledge, 2009.*

Lambek, Michael, ed. A Reader in the Anthropology of Religion. *Oxford: Blackwell Publishing, 2002.*

Pals, Daniel L. Introducing Religion. *New York: Oxford University Press, 2009.*

Rodriguez, Hillary and John S. Harding. Introduction to the Study of Religion. *New York: Routledge, 2009.*

VIDEO RECORDINGS:

Elda Hartley, producer and director. World Religions, Many Paths, *Vol. 1, videorecording/199 minutes. Westport, Connecticut: Hartley Film Foundation, 2006.*

North South Productions for Channel Four Schools Television. Animism: Living in the Dreamtime, *1 videodisc (15 minutes). Lawrenceville, New Jersey: Cambridge Educational, 2004.*

NOTES:

[1] *Mircea Eliade,* The Quest *(Chicago: The University of Chicago Press, 1969), 1.*

[2] *Paul Tillich,* What is Religion? *(New York: Harper and Row, 1969), 10.*

[3] *Karl Rahner and others, eds.,* Sacramentum Mundi *(New York: Herder and Herder, 1970), 247.*

[4] *James Thrower,* Religion *(Washington, D.C.: Georgetown University Press, 1999), 51.*

NOTES, continued:

[5] *James Forsyth*, Psychological Theories of Religion *(Upper Saddle River, New Jersey: Prentice Hall, 2003), 219.*

[6] *Ibid., 245.*

[7] *Ibid., 247.*

[8] *Thrower, 1.*

[9] *Eric Sharpe*, Understanding Religion *(London: Duckworth, 1983), 87.*

[10] *See Roger Haight*, An Alternate Vision: An Interpretation of Liberation Theology *(New York: Paulist Press, 1985); Peter Beyer*, Religions in Global Society *(New York: Routledge, 2006); K.R. Dark, ed.,* Religions and International Relations *(New York: Macmillan, 2000).*

[11] *Tillich, 165.*

[12] *See Lance deHaven-Smith*, Foundations of Representative Democracy *(New York: Peter Lang, 1999), 187ff; Robert A. Segal, ed.,* The Blackwell Companion to the Study of Religion *(Chichester: Wiley-Blackwell, 2009).*

[13] *Forsyth, 150.*

[14] *Thrower, 60.*

[15] *Victor Frankl*, Man's Search for Meaning *(New York: Washington Square Press, 1963).*

[16] *Forsyth, 245.*

[17] *Karl Rahner*, Foundations of Christian Theology *(New York: Crossroad, 1987), 126ff.*

[18] *David Tracy*, Blessed Rage for Order *(New York: Seabury Press, 1975), 119.*

[19] *See Catholicism.org, Feb. 11, 2010.*

[20] *See Bron Taylor*, Dark Green Religion *(Berkeley: University of California Press, 2010).*

[21] *Elizabeth Tetlow*, Women and Ministry in the New Testament *(Mahwah, New Jersey: Paulist Press, 1989), 5ff.*

[22] *See Li-Hsiang Lisa Rosenlee*, Confucianism and Women *(Albany, NY: SUNY Press, 2006).*

Hinduism

Hinduism: A Personal Encounter

It was a warm June night in Varanasi, India. Our guide helped us into a large flatboat and rowed us out into the Ganges River. We sat, surrounded by boats and facing the shore, where we saw thousands of Hindu pilgrims watching a celebration. Seven altars faced the water, each with a priest dressed in colorful robes and swinging large flaming urns. Flowers and candles were everywhere, and the chanting of hymns mixed with the ringing of bells. On the far ends of the altars, large fires burned as the bodies of the deceased were cremated. At one point, all the pilgrims in the boats lit candles in small baskets filled with saffron flowers and set them adrift on the river. Thousands of these small lights floated downstream. This celebration of Mother Ganges is performed every evening, an amazing ritual of the oldest religion of the world, Hinduism.

>> *See Benares, Ganga Aarto Ceremony:*
http://www.youtube.com/watch?v=4O8BfbtQKg0

Hinduism, a religion more than seven thousand years old, is extremely complex. It is not only a religion; it is also a cultural tradition and a way of life.[1] Hinduism, or more properly the "Sanatana Dharma" (the Eternal Law), is unique as a religion because it has no single founder, organized creed, or single set of scriptures. Hinduism worships one absolute reality (*Brahma*) that can be experienced within each person (*atman* or soul), the union of the human with the divine. It is a religion with a seemingly endless list of gods and goddesses, multiple divisions and sects, and a broad spectrum of practices. Hinduism has been described as a religious "process," a developing tradition that has changed throughout many centuries and that continues to evolve.[2] With approximately 950 million followers, Hinduism is the third-largest religion after Christianity and Islam. It is the dominant religion in India, Nepal, Bali, and among the Tamils in Sri Lanka.

Approaching the study of Hinduism requires the ability to connect multiple elements into one holistic process of faith. In the following sections, we will consider the beginnings of Hinduism, its sacred texts, beliefs in many gods and goddesses, the origins and continuation of the caste system, and the **three paths of Hindu life**—knowledge, works, and devotion. Then, we will examine how modern-day Hindus apply their beliefs to the issues of ecology, peace, and women's issues.

THE BEGINNING OF HINDUISM

>> *See* **The Story of India :**
   www.youtube.com/watch?v=MPOKyrnjpUA

The words "India" and "Hindu" are derived from the Indus River, which flows near the western border of India. The people of India were designated as the people of the *al-Hind* or Indian subcontinent. Their religion was designated as Hinduism in the late eighteenth century by India's British Colonial rulers.

The story of this ancient religion begins in the Indus Valley in the lower Himalayas, an area that only became known to the Western world in the 1920s when British archaeologists arrived. Their excavations uncovered a vast ancient civilization, which included cities dating back to 2800

BCE. We know little about life in this civilization, but the unearthing of its many baths indicates concern for ritual cleansing, which is still so important to Hindus today. The discovery of many seals and figurines seems to indicate a veneration of images, especially goddesses, a concern for sexuality, and a veneration of animals. This culture also seemed preoccupied with the suffering aspect of life (**dukkha**), an emphasis on asceticism by wandering mendicants, and currents of atheism, all of which would influence Hinduism and especially the later development of Buddhism. More recent excavations have moved the beginning of this civilization back several more millennia.

One theory regarding the origin of this civilization—a highly controversial one—is that a nomadic Aryan people, possibly from what is now southern Russia, began to move into the Indus Valley around 2000 BCE.[3] The word **Aryan** means "the noble ones." By the time the Aryans arrived in the Indus Valley, most likely from migration, they were an advanced culture. They brought with them horses, chariots, bronze tools, weapons, and their language, **Sanskrit**, which would become the classical language of Hinduism.

The Aryans also brought with them a rich religious tradition, which would ultimately blend with that of the Indus culture to form the foundations of early Hinduism. The Aryans had developed sacred chants, fire ceremonies, and elaborate sacrifices of animals led by priests and accompanied by a fermented drink called soma.

THE SACRED TEXTS

The development of Hinduism can be traced through its sacred texts. Eventually, the Hindu writings would be divided into **Sruti**, the writings that were heard, and the **Smriti**, the writings that were remembered. The Sruti are infallible revelations and include the Vedas (hymns) and Upanishads (scriptures). The Smriti, which are not seen as infallible revelation, are the written Codes of Conduct, such as the *Laws of Manu*, and the

three epics, *Mahabharata*, *Ramayana*, and Bhagavad Gita, the latter being the most influential. Finally, there is a collection called the Puranas,
which consists of devotional literature that features the gods Vishnu and
Shiva, and a collection of Sutras, which are short sacred texts that can be
easily memorized and recited.

The Vedas - The Vedas (1200–900 BCE) are an extraordinary collection of hymns that were sung at the great sacrifices. Composed between
1200 and 900 BCE, they were highly creative, filled with emotion and
imagination. The Vedas are part of the Sruti, the highest form of revelation. Some Hindu traditions believe that these revelations came from an
eternal source and were seen or envisaged by the original reciters. These
revelations were first recorded in an oral tradition and were later written
down.[4] Besides the Vedas, the Sruti include the **Brahmana** (ritual books),
the **Aranyakas** (forest meditations), and the **Upanishads**, the most important revelation in the collection, that speak of a universal spirit (Brahman) and of an individual Self (Atman).

>> *Listen and watch the chanting of* **Mantra Pushpam:**
Vedic Hymns in Sanskrit
http://www.youtube.com/watch?v=-yS-Jky997Y
as it explains how water is the basis of the universe.

The oldest and most significant of the four Veda collections is the **Rig
Veda**, which alone has one thousand hymns. In these hymns, more than
thirty-three deities (**devas**) are venerated. Indra, the god of the sky and
warfare, seems to be the most popular god. The people felt close to him as
they sang: "May Indra come to us with life and friendship."[5] The disciples
often called on Indra for prosperity and for strength in battle:

*"Strong God, the riches which thy hands have from days of old
have perished not nor are wasted. Splendid are thou, O Indra, wise,
unbending; strengthen us with might. O Lord of power."*[6]

Besides Indra, Varuna was honored as the god of order and the one to go to for forgiveness. Rudra was a primitive, wild god, whose characteristics reappear in the later Shiva. Agni, the god of fire, gave the people power for cooking, heating, light, and cremation. Vishnu, who will later become one of the main gods, is only given a mention at this point. Surya is the sun god and Usas, the goddess of dawn. Here is an example:

"Munificent dawn awakens men curled up asleep; one for enjoyment, another for devotion, and another to seek for wealth; they who could scarcely see, now see clearly. All living beings are now awakened."[7]

These amazing Vedic hymns seem to have been developed for offering elaborate sacrifices to the gods. The devas (gods) were praised extensively and requests were made for wealth, many cattle, good crops, and victory in battle.

Imagine the ancient scene of people standing around a huge blazing fire near an enormous altar, surrounded by priests dressed in elaborate robes. **Soma**, an intoxicating drink, is passed around and poured on the altar. Throughout the night hundreds of valuable animals are sacrificed to the gods. In this great valley of the Himalayas, these people experience the infinity of the sky and the frightening power of thunder and lightning on a summer night. They called this power Indra, the sky god. As the sun goes down and the sky fills with countless stars, a chant rises up from the people.

"O COME you hither, sit you down: to Indra sing you forth, your song, companions, bringing hymns of praise. To him the richest of the rich, the Lord of treasures, excellent Indra, with Soma juice outpoured. May he stand by us in our need and in abundance for our wealth."[8]

The sounds of the hymns ring through the valley throughout the night. In the morning, the power and beauty of the sunrise over the mountains is experienced, and the people sing to Usas, the goddess of the dawn: "We behold her, daughter of sky, youthful, robed in white, driving forth the darkness."[9] The people return to their homes secure under the protection of their gods and hopeful for more prosperous lives and victories over their enemies.

Gradually, these sacrifices and the attendant formulas, through which the cosmic and the human were thought to be sustained, became identified with the ultimate reality itself. That ultimate reality was called **Brahman**. The Rig Veda announces this revelation of the divine: "That One Thing, breathless, breathed by its own nature; apart from it was nothing whatsoever."[10] The Vedas also explored the mystery of creation. Conceptions of the deity as Creator of all things, the Lord of all Creatures, began to emerge. The people asked: "Whence comes this creation?" But there were no easy answers: "Whether he brought it forth or not, He who beholds it from the highest point of heaven, Only he knows, or perhaps even he does not know."[11] One creation story recounts the dismemberment of a cosmic person, Purusha. The pieces are made into the castes, the elements of the cosmos, and the gods.

> *"His mouth was the Brahman [caste - priests], his arms were the Rajanaya [Ksatriya caste - warriors], his thighs the Vaishya [caste - merchants and farmers]; from his feet the Shudra [caste - servants] was born. The moon was born from his mind; from his eye the sun was born; from his mouth both Indra and Agni [fire]; from his breath Vayu [wind] was born. From his navel arose the air; from his head the heaven evolved; from his feet the earth; the [four] directions from his ear.*[12]

As the Indo-Aryan culture moved south to the Ganges area, the religion began to move beyond the Vedas and preoccupation with this world and material things. New ultimate questions would arise.

The Upanishads - The Vedic period ends around the time that the Axial period began (800–200 BCE). This era saw the explosion of philosophical and religious thought: Greek philosophy, Jewish prophets, Buddhism and Jainism in India, Zoroastrianism in Persia, and Confucianism and Taoism in China.

>> *Experience the Upanishads' description of God*
http://www.youtube.com/watch?v=mTfEXfA7OAs&feature=related
and consider the connection of this ancient scripture in relation
to our modern world.

It was during the Axial period that the Upanishads appeared in India. The **Upanishads** (800–500 BCE) are a group of writings that expressed the mystical and intuitive experiences that were advancing the Hindu tradition. Simply praising the gods and asking for things through sacrifices were no longer enough. Hinduism now had new ultimate questions, such as: "Who is God?" These ancient peoples were coming to believe that God was beyond all things, and yet within all things. They declared:

> *"He transcends all things; and yet he is present in all things. He does not have a body or a mind, and yet he had created all bodies and minds. He does not have form or name, and yet he is the source of all elements—space, fire, air, water and earth..."*[13]

They asked questions that are still asked by many today: "Is this all there is? Is there more to life than material things and pleasures? Is there a deeper meaning and purpose to life?" They answered:

> *"When God created humans, he gave them senses to grasp external objects and happenings. But wise people turn the senses inward, to perceive the soul. Foolish people chase outward pleasures, and so they fall into the trap of death. Wise people know that such pleasures are fleeting and they ignore them, wanting only joy that is eternal."*[14]

Suddenly the impermanence of things, the constant changes in everything, and the prevalence of suffering and death became more apparent.

Whereas the Vedic period was more focused on this life and viewed death as simply going to the heavens or perhaps being recycled into the elements, during the Axial period there were new questions. Is death really the end? Are humans caught in an endless cycle of birth, life, death, and rebirth?

The notion of **reincarnation**, which the Greeks had explored philo-sophically, now entered the religious realm and became important for the Hindu tradition and their understanding of rebirth. The Upanishad points out: "After death the soul may return to the mother and obtain a new body. The soul may become a tree or a plant."[15]

Another important question arose. What is the relation between the moral life and the afterlife? Are there simply rewards and punishments, or does one's behavior (**karma**) determine what rebirth will be? New an-swers to the perennial question of evil appeared. Perhaps people suffer because they bring it on themselves. People can be masters of their own fate; they are no longer at the mercy of the gods. The gods may well assist them with grace, but people were able to make their own decisions about their existence and were willing to accept the consequences. The notion of **karma** began to appear as a key element in the Hindu tradition:

> *As people act, so they become. If their actions are good, they be-come good; if their actions are bad, they become bad. Good deeds purify those who perform them; bad deeds pollute those who per-form them....Thus we may say that we are what we desire."*[16]

Another query arose regarding reality itself. If Brahman is the ultimate reality, what is real about the world? Is the world around me and all that I experience a mere illusion (**Maya**)? Perhaps salvation lies in liberation from this world with all its limitations and in seeking union with **Brah-man**, the Absolute. **Moksha**, or liberation—the release from the worldly limitations of being an individual—becomes the life goal.

And finally, new consideration was given to how to relate with Brah-man in this life. The Vedic tradition had developed a relationship of praise and petition with the gods (devas). During the Axial period there was a move toward a more personal relationship with Brahman.

> *"Those who find God within themselves and find God within others will have no need to fear. Those who see themselves as one with all things and know the unity of all things will have no sorrow."*[17]

In the more than one hundred writings of the Upanishads the relationship of an individual with God became more intimate. The beyond could be experienced as within, the transcendent as immanent. Brahman, the only absolute reality, was indeed within everything that was observed. At the same time, Brahman remained a mystery, beyond conceptualization, sight, hearing, and yet the very basis for all these human powers.

"When words try to approach God, they are forced to turn back. Human thought can never reach God. Yet human beings can know the bliss of God, and thereby be freed from all fear."[18]

More concern was then given to the inner self, the Atman-Brahman, or the inner self of all beings. Brahman is the absolute reality that supports the visible world, and **atman** is the inner experience that supports the absolute within each person.

"When people find and recognize the soul within themselves, they realize that the soul is the Creator of all things, the author of the universe itself."[19]

The tradition was seeing the union of the human with the divine. *"Thou art that"* became the breakthrough mantra of the time, reflecting awareness that the deepest inner self is identified with the Ultimate Reality. This Ultimate Reality is now understood to be Brahman, and throughout the Upanishads the people search to understand the Ultimate Reality, and how they can be one with that reality. This ancient prayer could well be contemporary in its expression:

"Cause me to pass from the unreal to the real, from darkness to light, from death to immortality."[20]

There are two key Hindu epics that deal with the human search to be united with the divine: the Ramayana (story of the Rama) and the *Mahabharata* (Great Epic of India). The Bhagavad Gita (sacred song) is the famous episode that appears in the *Mahabharata*.

The Ramayana - Gandhi praised *Ramayana* (ca. 800 BCE) as the greatest religious literature in history. Today, its stories are still read and sung by many Hindus. It is the epic story of Rama, who is exiled by his father, and who struggled to rescue his wife Sita, who has been captured by the demon king, Ravana. Rama kills Ravana, rescues Sita, and returns home where he becomes the ideal king. Some think this is an allegory about the search for the absolute reality, where Rama, a god, rescues Sita, the soul, from Ravana, the body, and gives the soul liberation. The peace that comes when God reigns is magnificent:

> *"Ten thousand years Ayodhyá, blest*
> *Ráma's rule with peace and rest.*
> *No widow mourned her murdered mate,*
> *No house was ever desolate.*
> *The happy land no diseases knew.*
> *The flocks and herds increased and grew.*
> *The earth her kindly fruits supplied,*
> *No harvest failed, no children died.*
> *Unknown were want, disease, and crime:*
> *So calm, so happy was the time."*[21]

The stories celebrate many human virtues and present admirable role models for disciples. There is the obedience of Rama to his father and his constant generosity and courage; there is Sita's dedicated fidelity to her husband Rama, even in times of distress. There is the hope and delight in seeing evil beings defeated throughout the stories. There is the belief that in telling these inspiring stories the reader will be turned from sin and evil.[22]

The Mahabharata - The *Mahabharata* (ca. 900 BCE oral form) is treasured literature for the Indian people. This great epic of one hundred thousand verses contains many tales of holy rulers, courageous warriors, devoted wives, and fervent ascetics of the forest. It is literature that developed during a period of eight hundred years (400 BCE–400 CE) and was influenced by bards, warriors, teachers, and devoted followers. It is a book that reveals the ideals of law, morality, leadership, and social life. In its accounts of wars, one sees metaphors where evil is constantly defeated by good, and the higher powers of the person prevail over the lower. Readers are moved to be faithful to their loved ones and friends and to always set their hearts on that which is good and noble.

The Bhagavad Gita - The **Bhagavad Gita** (ca. 100 CE), part of the *Mahabharata*, seems to be the favorite religious writing among Hindus. Although it is not part of the Sriti, the revealed literature, it has a place of its own and bears more influence than the Vedas or earlier Upanishads. Here, ritual and searching no longer hold center position. Now, there is an intense awareness that God and humans dwell within each other. Salvation comes through love and devotion to that same God: "They, who with devotion worship me, are in me and I in them."[23]

The Bhagavad Gita frames its revelation of union with a personal God in an epic account of a battle over the throne. In the epic, Krishna advises Arjuna, a prince who accompanies him into battle, how he should fight. Krishna explains that Arjuna must fight because: (1) it is the duty of his state in life as a warrior; (2) his obligations must not be carried out for personal success or material gain, but simply because it is the right thing to do; and (3) God considers humans to be his friends and sends **avatars**, divine figures in human form like himself, Krishna, for protection and to reveal the truth.

The central question of how to be saved is clearly addressed by Krishna: *"Direct your mind to me alone. Be detached from praise or blame, be disciplined, accept whatever happens, detached and then: That person is dear to me."*[24]

The epic ends with Arjuna being given the vision of God and professing his faith in the Infinite Lord of Gods. Krishna has taught Arjuna the ultimate lesson of life: devout love and offerings to God, with the help of divine grace, will attain eternal happiness.

In the Bhagavad Gita, the soul is considered to be "God's spirit," which dwells in every living being and is described as immortal: "Though every body is finite in size, the soul is infinite. Though every body dies, the soul is immortal."[25] The soul is considered to be present in all living things, which results in a universal equality: "The soul is equally present in the priest, outcast, elephant and cow. This should influence you to treat all living things with equal respect."[26] Inner peace is essential for a happy soul and comes from detachment: "When you free yourself from all attachments and are indifferent to success or failure, then do you experience inner peace."[27]

>> *Visit the* **Heart of Hinduism** *Website*
http://hinduism.iskcon.org/concepts/109.htm,
to learn more about the different path Hindus may take
to the ultimate goal of eternal moksha.

The Puranas - Several centuries before and after the beginnings of the Christian era, the stage was set for new developments in Hinduism. Temples were being built throughout India, images of the gods and goddesses became popular, and **bhakti**, a strong commitment to devotion to God, developed. This form of Hinduism prevails today. During this period, two deities became prominent, **Shiva** and **Vishnu**. Likewise, two principal religious sects developed: the **Shaivites** who worship Shiva and the **Vaishnavites** who worship the god Vishnu. At the same time, a feminine principle and energy, **Shakti**, became associated with these gods. In contrast, **Shaktism**, another sect of Hinduism that worships Shakti as the dynamic feminine aspect of the Supreme Divine, was established. New scriptures were needed to express these developments, and the Puranas became the most important of these.

The **Puranas** (500 BCE–1500 CE) were written throughout many centuries. They repeated many of the traditions of the past, but added new mythologies, local practices, and instructions on how to practice this growing religion. The Puranas focused on worship of Vishnu, Shiva, and Brahma. Shiva has a female consort, Devi, sometimes known as Durga, Kali, or Parvati. The teaching of **avatars**, deities appearing as living beings, appears in the Puranas, as well as new and endearing stories about Krishna, including his playful times with milkmaids and his romantic love for Radha. The four basic aims of life are listed: carry out the duties of one's state in life, seek wealth, pursue pleasure, and strive for liberation.

Other writings include the **Agamas** (500 BCE–1800 CE), which are the fundamental philosophical and traditional base of present-day Hinduism. The Agamas, including Tantras, Mantras, and Yantras, deal with the philosophy of deity worship and give instructions for practicing yoga meditation, building temples and creating figures of deities for worship, and celebrating rituals. The **Tantras** deal with symbolism, the sacramental system, and further teaching on yoga. And finally, many more hymns were composed, including a large number in the great Tamil tradition in southern India and northeast Sri Lanka.

GODS AND GODDESSES

Every Wednesday evening a Hindu community gathers for a worship service at their temple near my university in Cincinnati, Ohio. Attending a service is an opportunity to observe the Hindu religion in action.

As my class and I arrive at the temple and mount the stairs, we notice many pairs of shoes outside the door, a signal that we are to remove our own shoes before entering this sacred place. A young Hindu priest greets us and invites us into a large, carpeted area. He asks us to ring the bell overhead to announce that we are there, prepared to pray along with the approximately one hundred Hindus present: the women in colorful saris, the men in loose white trousers and long silken jackets.

Sitting down on the carpet, we observe at the front of the room a large area with many statues. One has the head of an elephant; another has a monkey's face. A large crystal figure of OM, resembling the shape of a three next to a zero (3 0), sits in the middle of the grouping. The statues are of both male and female figures, and they are elaborately dressed. Worshipers are praying before these statues, and some place food before the figures.

The priest explains that all the worship in this temple is of the one Ultimate Reality, Brahman. The people are not worshiping idols; rather, they believe that through the intercession of the devas (deities) these images represent, they can commune with the deity and receive blessings. He gives an example: When you talk into your phone, you are not talking to the phone, but merely using it as a means to talk with someone else. The Hindus believe that the Divine is present everywhere and that these figures are symbols of the divine presence. Each figure represents some power of the Divine, which the people can invoke, depending on their needs.

The priest starts with the crystal figure in the middle. He explains that the figure OM (sometimes spelled as AUM) is thought to be the sound made when the universe was created, the sound on the lips of the faithful before every prayer. He points out that in sound, it is similar to the Christian Amen and the Islamic Amin.

Next to the OM figure stands the trinity of statues representing the three key gods of Hinduism: Brahma the Creator, Vishnu the Preserver, and Shiva the Destroyer. The priest gave us a way to remember this trinity in terms of the word god: G = Brahma the Generator, O = Vishnu the Orderer, and D = Shiva the Destroyer.

We had many questions about these statues: What is the significance of these statues? Who are they, and why are they so ornately costumed? Why do some of the statues have animal shapes? Why are some so fearsome?

Lord Krishna - On the far left is a statue of Krishna, one of the deities most favored by Hindus and a main character in the epic Bhagavad

Gita. Krishna is an avatar of the god Vishnu, who is one of the main gods of Hinduism. (The term Avatar refers to a deity who has descended into bodily form.) The stories about these two, as well as prayers to them, can help the believers better relate to God in loving devotion.

Ganesha - The figure next to Krishna has an elephant head. The priest explains that this is Ganesha, the son of Shiva. When someone needs help in an exam or to succeed in a job, Ganesha is the one to go to because he is so clever. He is pictured with a human body and an elephant head. There are many stories about how he received his head. One recounts that Shiva cut his son's head off thinking Ganesha was a stranger in the house. In the aftermath, Shiva sent his servants to the woods, where he told them to bring back the head of any sleeping creature facing north (a direction associated with wisdom). They came upon an elephant and sacrificed it, and Shiva grafted the elephant's head on his son's body and revived the boy. Ganesha is often pictured as having four arms and standing on a mouse, symbolic of his power to get rid of bothersome things; he is considered the Remover of Obstacles. Hindu children love to hear stories about Ganesha.

Lord Shiva - There is also a huge statue of a dancing man, whom the priest describes as a major deity, Shiva. This particular Hindu community practices Shaivism, which means that they focus their worship on Shiva rather than Vishnu. Shiva is believed to have developed from Rudra in Veda times, a god who is fierce, destructive, and frightening. He carries a sharp trident, and worshipers ask him to ward off evil. At times, he is portrayed with ashes, showing his role as the one who presides at cremations. Shiva is worshiped by the elderly to ward off death or to destroy the fear of death. He can also be benign and bring happiness. Shiva can also be shown in the lotus position as a prayerful ascetic and is worshiped by those who wish to do well at meditation. Commonly, Shiva is surrounded by fire, holding his leg up dancing the Tandava. His wife, also the source of his energy, is Parvati.

>> **Listen to Om Namah Shivaya, an adoration of Shiva**
http://www.youtube.com/watch?v=w46gmDfUaQ0&feature=related
and explore the meaning of this mantra, particularly to Shaivites.

Mother Durga - *The statue of a beautiful woman dressed in regal garments and sitting on a lion represents Mother Durga. With her eight hands, she is a symbol of the great power of woman. Durga is a goddess to whom one can turn for courage and strength during difficult times.*

Rama and Sita - *A pair of statues represents a splendid couple dressed in royal garments and wearing crowns. We are told that they are Rama and Sita, and they represent the ideal man and woman. They are the reincarnations of Vishnu and his wife Lakshmi, the goddess of light, beauty, wealth, and good fortune, who plays a special role as the mediator between her husband Lord Vishnu and his worldly devotees. Rama is the model son and husband, the just ruler, and enemy of evildoers. Rama is worshiped for those who seek harmony in their families, peace, or help in dealing with an evil person.*

Hanuman - *Finally, the priest points out Hanuman, who has a human body and the face of a monkey. In the great epic, Ramayana, Hanuman is the devoted and talented chief of the army. He is able to conquer all situations and bring many blessings. Hanuman is thought to have been given the monkey face to show how leaders can be led astray by material gain. Not so, of course, for Hanuman. He is always faithful to his duties.*

>> Observe ceremonial worship of a deity using
a festal lamp during the Hindu ritual of aarti,
Om jai jagdish hare:
http://www.youtube.com/watch?v=3BPHI_120gw&feature=related

Brahman - *The priest then explained, as we read earlier in the Upanishads, that Brahman is seen as present in all that exists, and is indeed the foundation and source of all that exists. Brahman is beyond comprehension. Simply put: "He is," which calls to mind the Hebrew notion of God as "I am." Brahman is the divine Creator, who depends on nothing else for existence, the ruler of creation, who transcends creation and yet is within all of creation. Brahman is the savior and liberator who brings humans to immortality.[28]*

With the Hindu pantheon before us explained, the priest then moved off to the side and led the community in song. For some time, accompanied by music, the community chanted the beautiful hymns of their faith, most seeming to know the songs by heart. When the singing ended, graceful dances were performed. When finished, many went to the altar and ate the offerings of food blessed by the gods and goddesses. The Hindu worship service was over, and the community gathered outside the worship area to socialize.

TRADITIONS, CLASSIFICATIONS, AND PATHS TO LIBERATION

Hindus often view the classifications of their religious practices as revealed, which places these traditions firmly in the culture. We saw earlier how creation resulted from the dismemberment of a cosmic being into four classes or castes: priests, warriors, farmers and merchants, and servants. Early Puranas spoke of the four aims of life: following moral law, gaining material goods, enjoyment, and liberation. Even today, the three goals for many Hindus are: duty (dharma), prosperity (artha), and Kama (sensual pleasure).[29] The **Laws of Manu**, the extensive library of Hindu legislation, speak of the four traditional stages of life: the student; the married householder dedicated to the duties of family, enjoyment of life, and the acquisition of material wealth; the retired person dedicated to the spiritual life in the forest; and the elderly person seeking liberation.[30] The Upanishads speak of four similar stages of life. These structures continue to deeply affect the personal and social lives of Hindus today.

In modern India we see an excellent educational system, and Indians throughout the world place great emphasis on schooling in the Hindu tradition. There is still a serious and often strict selection system for choosing marriage partners, along with the celebration of protracted and expensive weddings. Strong family units are dedicated to their children, to material gain, and to the celebration of life. One still sees elderly sadhus (ascetics) along the banks of the Ganges River—men who have set aside everything to dedicate themselves to the achievement of moksha (liberation) and to the mentoring of others.

Modern India experiences a tension between these traditional structures and those evolving from its modern industrial, technological, political, and global advances of the last century. As we shall see when we address the engagement of Hinduism with social issues, it is challenging to participate vigorously in industrialization and prosperity, yet still be committed to environmental concerns. It is difficult to promote the sacredness of all and ahimsa (nonviolence) while engaged in a nuclear arms race. It is not easy to balance the hierarchical and patriarchal structures of Brahmanism with modern democracy or the contemporary women's movement.

The Caste System - From the early Vedic period, orders of classes existed in Hinduism. This resulted in a hierarchy of social classes (varnas) among Hindus, as well as an elaborate structure of thousands of subdivisions (jatis) based on locality of birth, occupation, and local language. Many castes have their own religious myths and practices. Castes are a strong factor in deciding whom one can marry, and castes have their own political parties and religious practices. In a typical Indian village today, there is a complicated division among the people according to birth, profession, and other factors. Many of these dividing factors are outside the realm of caste, which renders the entire caste system extremely difficult to define. One can belong to a traditional caste, a professional caste, and sub-castes based upon his or her locale, language, kinship, and even specific rituals performed. Today, it is noted that not only religion but also economics and politics have played roles in the development of the caste system.[31] Therefore, though one may think of castes—"watertight compartments, where people are rigidly locked into from birth"[32]—as a stereotype, it also needs to be noted that in the modern cities of India these divisions are not so evident in social or corporate life.

Added to the layers of castes is a large group with a population of more than 300 million who belong to no caste at all (outcasts). Somewhere in their lineage, caste rules were broken, and an individual or family was thrown out of a caste. From that point on, everyone born into the group was an outcast with no way to escape this fate. Such people have traditionally been called the "untouchables." They were excluded from the temples, were given the lowliest jobs (cleaning toilets, cremating the dead, sweeping the streets), and were strictly segregated. The cause of the untouchables was championed by Gandhi, who called them *harijan*, or children of God. Eventually, untouchability was outlawed in India, but nevertheless it still remains a social reality. There are still many examples of discrimination against untouchables, now called Dalits.[34]

In spite of the prejudice against them, the large number of Dalits in India led to their becoming a strong political force. K.R. Narayan (1920–2005), born a Dalit, became president of India in 1997. Kumari Mayawati (b. 1956), also a Dalit, was elected as chief minister of Uttar Pradesh, one of India's most densely populated states. She was reelected, serving a total of four terms.

>> *Watch the video India:* **The Dalit Story**
http://www.youtube.com/watch?v=G1l-x7708zY **to learn more about the caste system in India and its impact on the Dalit people.**

Paths to Liberation - In the Bhagavad Gita, Krishna teaches that the **Three Paths to Liberation** are: knowledge, works, and devotion. This model in many ways synthesizes the development of Hinduism and holds together its key elements. While all of these paths are usually followed by Hindus, different groups and even individuals may emphasize one or the other as being more important.

• *The Path of Knowledge*

Many Hindus look to the Upanishads for the basis of their thinking and spirituality. They learn from the scriptures through gurus, meditation, and yoga. The **guru** appears early on in Hinduism as the teacher who explains scriptures to students. (The word "guru" is derived from two words: "ru," which means light, and "gu," which means darkness. The guru brings light to the darkness.) In time, this role evolved to be a spiritual guide. Later, gurus initiated disciples into the religious orders and taught them the chants. The guru became qualified for these roles by spending years studying and learning the scriptures, as well as practicing yoga and meditation, thereby becoming detached from all things.

The guru stands in the place of Brahma, creatively leading, instructing, and inspiring. The guru reveals the secret insights of the tradition, designs special mantras for the disciple, and assists the disciple in discovering the truth that dwells within the self.

The Upanishads, an important guide for Hindu spirituality, are concerned with ultimate questions about the basis of existence and how everything is connected. These texts provide a method whereby one can look into the reality of self and all reality as "manifestations of the absolute."[35] In order to encounter Brahman, one must go inward. Led by a guru as teacher and the disciplined reflection of meditation, each person is called to separate the true self from the false self, illusions from the truth, and the real from the unreal. In the Upanishads, a person is called to meditation on the sun at daybreak as a means of moving into the light of truth:

"Let a person meditate on the (Om) as He who sends warmth (the sun in the sky). When the sun rises it sings as for the sake of all creatures. When it rises it destroys the fear of darkness. He who knows this, is able to destroy the fear of darkness (ignorance)."[36]

Realizing the Atman - The Hindus often interpret the word **atman** to mean the self or the soul. While the body is mortal and taken in death, self or soul stands as the "deathless" ground of the person.[37] The soul comes from the creative God and has the capacity of being united with the Creator. The Ultimate Reality, Brahman, can be found within the very depth of the person. We might say that the Ultimate Person, who is the breath of life within the person, can be discovered deeply within the self. The atman is where the deep-down freedom from evil, old age, and death resides, and it is in discovering that depth that one discovers the mystery of Brahman, the mystery of the Supreme Person. Then the atman and Brahman are recognized as being One. The self and Brahman are then inseparable, though not identical.[38] Of course, this unity can only be achieved after long years of discipline and meditation.

Understanding Karma and Rebirth - Integral to pursuing the path of knowledge is attending to one's karma. **Karma** refers to one's actions and follows each being after death. The Upanishads put it this way: "As one acts and as one behaves, so one becomes."[39] Whereas dharma (the teaching) is what one is required to do, karma is what one actually does. The knowledge of karma is based on the understanding of cause and effect: our actions exert a definite influence on our being and lives as well. The results of our actions, either good or bad, can accumulate. It is karma that causes rebirth, and Hindus dread being caught up in an endless cycle of deaths and rebirths. Becoming one with Brahman is the way of eliminating karma altogether and being free of death and rebirth.

In Hindu traditional thought the notion of karma is linked with belief in the afterlife for the soul. Vague in the Vedas, this notion was clarified in the Upanishads to mean a cycle of life, death, and rebirth (samsara). Traditions vary as to what this means, ranging from being relegated to various kinds of hells where there is darkness and fire, to a number of heavens, or to a paradise with God. Some traditions speak of intermediate stopping places, thus enabling surviving family members to experience the spirits of their ancestors and others. None of the texts are clear on what happens after death or during the times in between rebirths. Nor is there much discussion in the texts about people remembering their past lives. Traditionally, only great gurus and leaders have access to such remembrances.

What is common is the belief that karma brings about a cycle of life, death, and rebirths, and one can only be free of this cycle if one is separated from karma (both good and bad) altogether and united to Brahman. Heaven, then, is not so much a reward for good karma, as it is being completely liberated from karma. Karma is finite. It clings to us and brings about rebirths. We are reborn because we died with unfulfilled desires.[40] Once liberated from our desires and the finite, a soul can become one with the Infinite.

Not only does the notion of karma motivate Hindus to live good moral lives; it also offers them an explanation for the problem of evil. Afflictions are understood as results of bad actions in this life or in another life. While there is a degree of fatalism here, the power of free will choices is still recognized. Patient bearing of one's afflictions can contribute to salvation, whereas resistance and complaining results in more negative karma.

Achieving Liberation - The notion of **moksha**, liberation, arises from the realization that simply fulfilling our desires is not sufficient in life. One must be liberated from desire, and this goal is connected with death and whether one must be reborn and die again. The Hindu tradition developed the notion that there can be an endless cycle of rebirths if one does not discipline desire and lead a virtuous life. Liberation means freedom from the finite: liberation from all that holds one back from being one with the Infinite, the Absolute, Brahman.

The goal here is to be liberated from evil, and from the suffering connected to evil. The desire for liberation arises from setbacks, tragedies, and failures—all those experiences which make us intensely aware of the limitations within the self and the world. For the Hindu there is the belief that there is the infinite beyond all this, and that each one of us has the capacity to reach out and join with the ultimate. The goal is to free oneself from the ignorance, the desires, and the evils that hold us back and move us toward rebirth and away from fulfillment.

The Upanishads say:

"Those who act without any reference to personal gain, and who take control over their actions, will in time discover God; and then they will know that all is one. Those who live in the service of God, will be freed from the process of cause and effect."[42]

Successfully navigating the path of knowledge gains one's immortality and freedom from the cycle of death and rebirth. As the Brahmana text states: "He who knows this achieves immortality."

Practicing Yoga - **Yoga** is the practice of physical exercises designed to gain self-control of the body, mind, will, senses, and spirit. In Hinduism, yoga supplies a path to achieve the goal of detaching, gaining liberation, and uniting with the divine.

The Bhagavad Gita advises:

"Learn about yoga, which frees one from bondage. Yoga is the eternal path. Steps along this path cannot be reversed, nor can effort in yoga be wasted. Even slight progress frees one from fear."[43]

The Yoga Sutras point to eight "limbs" or branches of yoga. **Hatha yoga**, the most familiar, attempts to balance mind and body via physical postures or **asanas**, purification practices, controlled breathing to increase the **prana** (life energy within the body), and the calming of the mind through relaxation and meditation. Hatha yoga works on all eight of the limbs, doing the first five simultaneously. The limbs are as follows: (1) external control (*yama*), which includes refraining from harming others, deceit, lying, stealing, or abusing sex; (2) internal control (*niyama*), which calls for serenity, purity, and seeking goals; (3) correct position (*asana*), which involves performing the various postures comfortably and without distraction; (4) control of breathing (*pranayama*), which is a key to centering self and gaining concentration; (5) control of senses and eliminating distractions from external objects (*pratyahara*); (6) concentration on some single point, such as a flame, the heart, or God (*dharana*); (7) meditation, through which the point of concentration fades away (*dhyana*); and (8) contemplation, whereby one enters into a state of freedom (*samadhi*).

A more advanced form of yoga is **Raja yoga**, which, primarily concerned with the mind, focuses on the last four limbs of yoga. Yogic practice also is concerned with awakening energy sources that exist along the spine. These spiritual energies come from centers called **chakras**. Under careful and skilled instruction, the rise in energy from the base of the spine to the crown of the head can be experienced by the practitioner. This is seen also as a path to mystical experience. Other Hindu branches of yoga include Karma yoga (union through action), Jnana yoga (path of knowledge), and Bhakti yoga (loving devotion).

• The Path of Works

In the Vedic period, the notion arose that people could participate with the work of the gods through sacrifices and rituals. Seldom do Hindus practice the elaborate ritual sacrifices of ancient times, but for special occasions, like the departure of a beloved priest or for an anniversary, they will gather around a ceremonial fire and chant sacred hymns. In place of an animal sacrifice, a ball of rice may be thrown into the flames. On pilgrimage, it is common to have elaborate fires for night worship services. Today, what is offered—be it water, food, fruit, or a flower—is not so important. What matters is that the offering be made with a good heart and with love.

Much more common are the acts of **puja**, the daily rituals in the Hindu home. We will discuss these rituals in more detail when we deal with devotion. The devout Hindu family will perform such puja several times a day at the family altar, perhaps using incense, flowers, readings of sacred texts, and offerings of food that would be blessed by the gods before eating. Food is sacred to the Hindu and is believed necessary to purify the mind, strengthen good concentration, and help in detachment from desires. Food holds a special place both in domestic and temple services.

Whether the work of worship be done in temple, as we discussed earlier, or at home, Hindus believe that regular worship brings the grace of God. Whatever type of worship that is done—whether it be singing, dancing, reading, attending services, private prayer, or going on pilgrimage—they fulfill the ancient dharma and bring God's assistance and redemption.[44]

Purifying the Mind – Hindu practice is aimed at purifying the mind, and purity of mind opens the mind to better understanding of self and the world. Such purity is believed to increase the capacity for devotion and meditation, both of which lead to union with God.[45]

Purity goes hand in hand with merit, and merit is of serious concern to many Hindus. Both purification and merit can be gained through worship at home or in temple, by reciting names of deities, singing, dancing, using mantras, going on pilgrimages, receiving blessings from priests, and bathing at home and in the sacred rivers to cleanse, preparing for a loving encounter with Brahman.

Renouncing Sin and Performing Penance – The Bhagavad Gita is an important resource for Hindu morality and for the standards to be followed in carrying out the duties connected with one's position in life. Delusion, which results in lack of judgment, is the basis for all sin. This leads to the chief vices: anger and greed. Desires are also the basis for sin because they drive people to excesses in sex, pleasures, and in acquiring an abundance of material things. One of the Puranas teaches: "The earth is upheld by the truthfulness of those who have subdued their passions and are never contaminated by desire, delusion and anger."[46]

The serious sins in Hindu morality are those that we find in most traditional religions: murder, stealing, and adultery. The avoidance of alcohol and avoidance of sinners are added to the list of Hindu sins. The *Laws of Manu* speak of all kinds of physical punishment for such sins, including diseases, deformities, and disabilities. The scriptures also describe various hells where sinners are punished for their sins.

Each person is thought to have three qualities: darkness, passion, and goodness. Darkness can draw the person toward such negatives as ignorance and laziness. Passion can lead the person to self-centeredness. Goodness can draw people toward the Divine. The proper path is to allow goodness to be the guide in all choices.

Certain virtues are valued highly in Hindu ethics: compassion, duty, non-injury, hospitality, truthfulness, and detachment.[47] Positive actions would be unselfish actions, such as carrying out one's duties according to one's station in life, and doing good deeds without a desire for reward and without fear of punishment for omitting such deeds. As we shall see in the sections on Hindus in Action, some choose to act heroically on behalf of the earth, peace, or the rights of women.

Intention and the use of conscience are very important in determining Hindu morality. A person is admonished to look within the self for norms, and at the same time be cautious not to act out of selfish motives. Specific moral norms often evolve out of particular sects, locations, or family groups, and this diversity has prevented Hindu morality from being unduly legalistic or rigid.

There are traditional works for getting rid of sin in a Hindu's life. The most common penance is fasting, cutting back on eating and drinking for an extended period of time. Another might make a vow to give up some pleasure or some form of entertainment or relaxation. Purification by washing in a sacred river, such as the Ganges, might also stand as a penance for sin. Giving gifts, going on a pilgrimage, spending some time in a forest retreat, giving alms to the poor, or throwing some item that symbolizes one's sins into a fire are means to gain atonement.

Performing sacramental rites of passage – In Hinduism, the stages of life from conception to cremation are marked by sacraments, sacrifices, and rituals known as **Samskara**. At birth the baby is ritually washed, sacred texts are whispered in the baby's ear, and OM is traced on the tongue with honey (*Jatakarman*). The naming ceremony (*Namakarana*) is especially important and is held the twelfth day after birth, with much singing and the establishment of the horoscope. There are rituals for the first time the baby is taken outside (*Nishkramana*), the first feeding with solid food (*Annaprashana*), and the first haircut (*Chudakarana*). For the boys, there is a sacred thread ceremony (*Upanayana*), done first at age eight as an introduction to the concept of Brahman and an initiation into formal education and repeated again every several years, accompanied by singing, dancing, and eating. The three layers of the thread represent his need for good speech, mind, and body.

>> *Be a guest at a traditional Krishna Hindu Wedding:*
http://www.youtube.com/watch?v=ehnDFerM0_8
Notice the ceremony, the participants, and the rituals.

>> *See Youtube,* **Hindu Wedding:**
http://www.youtube.com/watch?v=j42prjIQCdQ

In marriage (*Vivaha*), there is a special ceremony of betrothal and an elaborate weeklong celebration before the actual wedding ceremony. Marriages are usually arranged by the families according to caste and class background. In the higher castes, the bride is expected to bring a dowry or at least expensive gifts. Hindu families living in the Western world sometimes adapt to the wedding customs of the country where they live.[48]

There are elaborate death rites in Hinduism (*Antyeshti*). After death, there is a rite for washing and dressing the body. Divine hymns are chanted as the body is taken to the place of cremation. Cremation is not only thought to be hygienic; it is considered the common way of freeing the soul from the body. Cremations are carried out by those in the untouchable caste, but are presided over by the eldest son, with head shaven and wearing a white robe. The death rites must be done meticulously in order to assure that the deceased migrates correctly from family to the ancestors (*preta-karma*) and does not become a ghost (*bhuta*). Rituals are performed for twelve days after the cremation. At that time, it is thought that the deceased is with the forebears. After the rites are finished, the male relatives cut their hair, and the house is completely cleaned in order to eliminate the pollution of death. At the end of the death period, there is a special meal where food is provided for the family and for the spirit of the deceased as well.

>> *Watch* **Cremation**
*http://www.youtube.com/watch?v=QFNKdMSl2A0 **to learn
more about the cremation rituals of modern-day India.***

• The Path of Devotion

We have seen that the goal of Hinduism is loving union with the Ultimate Reality (Brahman) and that two of the paths to that goal are knowledge and works. The third path is devotion, which includes temple worship, home worship, and the celebration of feasts and pilgrimages. These aspects of Hindu devotion are for developing a loving and devout heart with which one can relate to the gods and to all others.

The erection of temples began in the fifth century CE. Temples were to be palaces where the gods and goddesses could hold audiences for their devotees. Temples were built to conform to the laws of the cosmos, with squares reflecting the divine and circles symbolizing time. An inner square housed Brahma, who was then encircled by the other gods. An outer circle was established for the people. Temples were aligned according to the astronomical directions and often included images of planets. There exists a wide variety of sizes and styles of Hindu temples, some simply roadside structures, others vast complexes of buildings.

>> **View dwelling places for the gods in the video The Hindu Temple:** *http://www.youtube.com/watch?v=Yiupwfu_hOk* **Notice the art and architecture of the Hindu temples of India.**

In the modern temples, ample room is often provided for processions, dancing, and other worship rituals. Usually when one enters, a rope attached to a bell is pulled to arouse the attention of the images of deities. The images of favorite gods and goddesses are dressed in colorful clothes, symbolic items such as jewelry, weapons, and scepters. Usually there is a canopy of honor over each image. Some members read or meditate, while others place small gifts of food before an image.

Worshiping at Home – The Hindu householder will get up at sunrise and speak the name of the favorite deity. The first *puja* will be to direct the eyes at the palm of the hand and then touch the floor and bow to the images of the deities on the family altar. These deities are usually dressed in elaborate outfits. Prayers and mantras that are traditional to one's sect are then recited, and thoughts turn to what efforts the person will make to be righteous and prosperous. The devotee then takes daily ablutions, recalling that water is the source and sustainer of life. Some will perform a water ritual to drive out evil. Others will recite mantras on a set of beads. Whatever is done is performed lovingly and devoutly in the belief that these actions will bring blessing and salvation.

>> *How to have a Puja at home*
   http://www.youtube.com/watch?v=Pbxlh8oRNWU

Celebrating Festivals – Hindu festivals can be exciting events, with thousands of worshipers, tall temple-like chariots, richly decorated elephants and horses, and throngs of sadhus (holy men) dressed in colorful robes. The signals to begin and finish are often given by the blowing of silver trumpets.

Feasts are determined by phases of the moon and are carefully determined by astrologers. The grace and merit to be received are connected to the propitious time for the celebration. In India there are a number of feasts celebrated nationally, but there are numerous other local feasts. Here are some of the main festivals:

Holi - This spring festival can last as long as sixteen days. On the main holiday, huge bonfires are lit, and colored powder and water are thrown into the air to celebrate the coming of Spring.

Navaratri (Nine Nights) - This festival is celebrated with the new moon (September–October) and is in honor of a number of the goddesses, including Durga and Sarasvati, the goddess of learning. The last day, dedicated to Lakshmi, the goddess of good fortune, celebrates the universal Divine Mother and is considered an auspicious time for taking on new ventures and paths of learning.

Dipavali (Necklace of Lights) - This is a five-day festival celebrated on the new moon (October–November). Houses are decorated, new clothes are worn, presents are exchanged, and there are special meals. The Hindu New Year celebration is often celebrated in mid-April, though dates vary in different provinces. This is a time for cleaning and decorating the house, taking a purifying bath, and making a fresh start in life.

Krishna Janmashtami and Ganesh Chaturthi - In July–August the birthday of Krishna is celebrated (**Krishna Janmashtami**) with singing and dancing throughout the night. In September, the favorite elephant-headed Ganesha is worshiped and petitioned for good fortune (**Ganesh Chaturthi**).[49]

Completing Pilgrimages – The belief that the Divine is in everything has led Hindus to consider many rivers, mountains, forests, and cities in India to be sacred. Indeed, India is spoken of as Mother India, and the Ganges as Mother Ganges. To visit these places on pilgrimage, to worship or to climb a sacred mountain, or to bathe in the waters of a holy river can be a means to liberation. There are special places designated for various gods and goddesses. The power of the Goddess Shakti is present in more than one hundred places; Shiva dwells in sixty-eight places and Vishnu in eight.

>> *Participate in the major celebration of Kumbh Mela-India*
http://www.youtube.com/watch?v=WqvzEjgYTFA&feature=related
celebrated once every twelve years to bathe to gain immortality.

Sacred texts point to ancient sites that are places of pilgrimage. The temples there are considered to be places where one can cross over from death to life. Near many of these temples are rivers, lakes, or pools where the pilgrims can cleanse themselves physically and spiritually. Rivers like the Ganges can wash away the sins of the pilgrim. Water from these sacred places is often taken home and used in local rituals. Certain mountains are considered to be dwelling places of the gods. For instance, Shiva is thought to live on Mount Kailash.

Key among the holy cities of pilgrimage is Varanasi, where Shiva resides in the Golden Temple. Tens of thousands of pilgrims flow into the city daily. There are constant funeral processions carrying bodies to be cremated. Varanasi is also a place for healing miracles, and it is the seat of Hindu learning with such institutions as Banaras Hindu University and the Sampurnanand Sanskrit University. In the southern city of Tirupati, there are marvelous temples, all of which manifest Vishnu. In Kerala, another famous place of pilgrimage, there is a famous wooden Krishna temple, which is known as a place for miraculous healings. The famous pilgrimage to Kumbh Mela is an amazing event where up to 7 million pilgrims gather every twelve years and celebrate for months.

The constant flow of thousands of pilgrims across India to sacred places is a strong symbol of the deep devotion and religious conviction of the Hindus. They are committed to eternal life with God and are devoted to continue the search for this goal.

MODERN MOVEMENTS IN HINDUISM

Modern India is quickly becoming one of the leading world nations. Its major religion, Hinduism, faces key challenges as India moves into this position. Hinduism has been the source of an ancient caste system, which puts a strain on India's modern emphasis on democracy and equality. A large percentage of India's people still live in abject poverty, many of them sleeping in the streets at night in the urban areas and millions living in substandard lean-tos in the rural areas. Hindu's eternal law will have to inspire a stronger sense of sharing, compassion, and generosity.

Hinduism has always had a strong tradition of ahimsa (nonviolence), as evidenced in the life of their great leader and liberator, Mahatma Gandhi. Now, with nuclear weapons, a growing military, and tensions with their neighbor, Muslim Pakistan, Hinduism in India will be challenged to keep this treasured value of nonviolence in the foreground.

India's economic success has introduced a new materialism and secularism into India. Hinduism, with its ancient traditions of simplicity, asceticism, and personal devotion, will be challenged here also. Although success in business and career are compatible with the Hindu tradition, greed and consumerism are not.

CONNECTING HINDUISM WITH WORLD ISSUES

Hinduism in India faces many social issues as its growing population experiences the greatest economic development in its history. Many of its states face severe poverty and food shortages. There are serious problems in the areas of education, health care, infant mortality, sanitation, waste management, care for the elderly, and others. As mentioned before, there is an ever-present tension between Pakistan and India, as well as between Muslims and Hindus within India.

Many Hindu leaders study their ancient tradition searching for values that can move their people to better address these issues. In temples across India and in other countries, Hindus conduct educational programs on social issues and encourage their people to become more involved in social action. In the following sections, we will discuss some involvement that Hindus are having in our focus areas: ecology, peace, and the women's movement.

Hindu Values and Ecology - India is said to have developed the world's largest environmental movement over the last few decades. The country's population has tripled during the last fifty years and grows by 70 million per year. This rapid growth has put tremendous pressure on the land and resources and has produced an enormous increase in pollution and waste. Moving rapidly into the industrial age has drawn many into urban areas in search of jobs. Some of these new residents live in comfort, but vast numbers live as refugees in shantytowns or sleep on the streets. In the countryside, where modernization and industrialization are not widespread, many live in poverty, often desperately searching for clean water and fuel for heating and cooking.

Many think that India is torn between two ideals. The first ideal is that of Mahatma Gandhi: religious, village-centered, and dedicated to self-sufficiency, simplicity, and nonviolence. The second ideal is that of Gandhi's successor Jawaharlal Nehru and many of Nehru's successors: secular, urban-centered, individualistic, materialistic, and caught up in an arms race.[50] Following the second ideal has led India to become a major national force, but at the same time has created a major environmental crisis.

Hindu religious leaders point out that Hinduism has vast scriptural resources for dealing with India's environmental problems. The tradition believes in the presence of the divinity within all things, and the resulting sacredness of the cosmos and the earth. The Hindu religion holds to the belief in the divinely ordained interconnection of all things and to the values of living in harmony with nature. Hindus also believe in ahimsa, or doing no harm, which is the basis for its teaching of nonviolence. Given these magnificent traditions, many religious leaders are baffled at the disconnect between these beliefs and every Hindu's real-world responsibilities to sustaining the earth and its resources. The leaders point out that many Hindus celebrate and sing the praises of their sacred mountains and rivers, but don't see the incongruity of allowing their mountains to be deforested or their rivers to serve as dumps and sewers. They can celebrate the divinity of the heavens and at the same time allow their air to be heavily polluted. They go to the sacred rivers to ritually cleanse themselves, and yet don't seem to notice that these rivers are seriously contaminated health hazards. Some explain that this disconnect is the result of the dualisms and world-denying aspects of Hinduism. Others indicate that many young Hindus embrace the world and use practical and realistic approaches to link Hinduism's deep respect for nature with the obvious needs for sustaining the earth and its resources today.

• *The Sacredness of Living Things*

It is clear in Hinduism that the world belongs to the Ultimate Reality (Brahman). Brahman is the supreme Creator, the source and goal of all life. In the Bhagavad Gita, Krishna says: "My invisible Spirit is the fountain of life whereby this universe has its being. All things have their life in the Life, and I am their beginning and end."[51] At the same time, Hinduism believes that all things are in the Divine and that the Divine is in all things.

> "All the worlds have their rest in me....I am the taste of living waters and the light of the sun and the moon. I am OM, the sacred word of the Vedas, sound in silence, heroism in people. I am the pure fragrance that comes from the earth and the brightness of fire. I am the life of all living beings....And I am

everlasting, the seed of eternal life. I am the intelligence of the intelligent. I am the beauty of the beautiful. I am the power of those who are strong."[52]

It is in the Hindu tradition that all living beings have a soul and that Brahman makes his abode in each soul. Nothing can have knowledge or action without the power of the Deity: "I am the seed of all things that are; and no being that moves or moves not can ever be without me."[53] Given this divine indwelling and power within all, every living thing is sacred and must be cherished as such.[54] In some of the Vedic traditions, the divine is portrayed as the Universal Goddess, and she is portrayed as both the transcendent and immanent forces in the universe.

• *The Interconnection of All Creation*

The Hindu tradition teaches that all creation comes from God and is interconnected. The Rig Veda views creation as a living organism that has come from God: "The moon was created from his mind, the sun from his eyes, Indra and fire from his mouth, the wind from his breath, the atmosphere issued from his navel, the heaven from his head, the earth from his feet, the quarters from his ears."[55] The Upanishads emphasize this same "interdependence of people and nature including deities, seers, fathers, animals."[56]

The earth itself is seen as "God's Body," and as the Mother who supports and sustains her children and who, therefore, deserves reverence. The Hindu dedication to the protection of cows (viewing cows as a source of life) is a symbol of the human obligation to protect nonhuman life, which has its origin in God. Creation is believed to be for the benefit of all, and therefore the rights of all living things must be protected. In the Bhagavad Gita, Krishna describes the "interweaving of the forces of nature," and speaks out for the common good: "let the wise person work unselfishly for the good of all the world."[57]

Krishna teaches that we must work in harmony with others because of our deep interconnection with God and all things. "And when a person sees that God in himself is the same God in all that is, he hurts not himself by hurting others."[58]

Hindu environmentalists speak of living in harmony with the trees of India, which provide so much benefit, including medicinal and healing effects to its people.[59]

• *The Protection of All Living Things*

Ahimsa (do no harm) is a rule of conduct that bars the killing or injuring of living beings and is practiced in a variety of ways by Hindus. For some, it is central to their way of life, especially those who renounce the world and live as ascetics. For others, it is an ideal with exceptions: priests are allowed to kill animals for sacrifice, and those in the military are permitted to kill in war. Gandhi was willing to admit that in his tradition there was justification for violence, but for him, ahimsa was eternal truth. He held that while our animal side is violent, the human spirit is nonviolent and that the person in touch with his or her spirit will be committed to nonviolence. Gandhi taught those who would be nonviolent that they must first confront their own fears and their greed. For Gandhi, ahimsa was more than the lack of violence; it included the force of love and compassion, as well as self-sacrifice for the sake of others. He also taught that nonviolence was not passive or cowardly, but was instead fearless and courageous, willing to risk all to protect all living things. Gandhi's vision has inspired many of the environmentalists in India, especially those in the Chipko movement, an organization devoted to nonviolent resistance to the destruction of forests.

>> *Learn more about the Hindu environmentalist* **Chandi Prasad Bhatt, Maker of India** *www.youtube.com/watch?v=tcwY04s_mlM and his grassroots efforts to protect the forests of India.*

Hindus in Action

Chandi Prasad Bhatt (b. 1934) – Bhatt was a farmer and a Hindu priest at a famous Shiva Temple in Gopeshwar. Bhatt began his work as a Gandhian environmentalist, defending the jobs, farmlands, and forests of the Himalayan mountain people. Bhatt, because of this standing, was able to found the **Chipko movement** and is considered to be the founder of the modern environmental movement in India. The Chipko movement began when Gaura Devi (1925–1991) organized twenty-seven women and girls from her village to protect trees from government deforestation. They held hands and surrounded the trees to protect them. The women saved 2,500 trees and reclaimed their forest rights. Devi led many protests against the Indian government's abuse of the forests.

Bhatt lives simply in Gandhi style with his wife and five children, but has been a formidable force in defending the local villagers against oppression and in training them in skills to make items from the local wood, teaching them to plant trees, and helping them to build walls to avoid mudslides.

>> *View* **The Real Avatar: Saving a Mountain***:*
http://www.youtube.com/watch?v=R4tuTFZ3wXQ
What is your reaction to mining on the sacred mountain?

Learn more of Baba Amte's unmatchable work
http://www.youtube.com/watch?v=zxyPmjtjnnY
and reflect on the causes Amte championed.

>> **Read about the turning point in Baba's life at**
http://mss.niya.org/people/baba3_amte.php
and contact the Niya aid foundation directly with your questions.

Baba Amte (1914–2008) – Baba Amte was born into a wealthy family and enjoyed the good life as a young man. Meeting a leper on the road one day changed his life forever. He walked away from his law practice and dedicated himself to the service of lepers. Amte had lived for a time in the ashram run by Mahatma Gandhi and was very much influenced by the Indian leader. He was joined in his work by his wife, Sadhana, and together they opened homes for the disabled, the blind, and the poor. Though crippled with a spinal disease later in life, Baba took on the causes of primitive groups and brought education and health care to their villages. He vigorously opposed the building of mega dams because of the destruction they brought to the poor people who were displaced by them, as well as the long-term changes they made to the environment. His health-care facilities, now led by his two sons and their wives, all of whom are doctors, include a university, hospitals, and model villages that teach simple living, self-sufficiency, sustainability, and compassion for the poor and disabled.

Veer Bhadra Mishra – Veer was a professor of hydraulic engineering at the Banaras Hindu University and knows a great deal about river pollution and purification. *Time* magazine declared him "Hero of the Planet" recipient in 1999 for his work related to the cleaning of the Ganges River. He has also served as High Priest at the Sankat Mochan Temple at Varanasi and is well aware of both the sacredness of the Ganges to the Hindu people and the damages done to the Ganges by pilgrims.

Veer knows that while the Ganges may be holy, it is not pure.

It is filled with chemical wastes, sewage, and even the remains of human corpses. The people who use the river often contract hepatitis, typhoid, or cholera. For many years he has been pleading with government authorities to clean up the river, but officials have been slow to react.

>> *Watch the interview with Professor Veer Bhadra Mishra*
http://www.youtube.com/watch?v=A29-eq78Wsk
about pollution in the river Ganges and social responsibility.

See also **The Sacred Balance-Science and Spirituality**
http://www.youtube.com/watch?v=7qTFzu-ArQk&feature=related
to see how spirituality and science interact.

Hindu Values and Peace - In Hinduism, traditions of both violence and nonviolence co-exist. There are ancient traditions in Hinduism that support violence. As we saw earlier, the Vedas describe creation in terms of a dismemberment of a cosmic person and sanction enormous sacrifices of animals. The King-warrior class has been given the right to use force in defending the social order. The epics of Hinduism often contain violent battles and the gods are often portrayed as warlike figures. The Bhagavad Gita is told in the context of a war, which is ultimately sanctioned as a duty. Within modern India, there are strong militant groups that have varying positions on national unity, the partitioning of Pakistan, and the presence of non-Hindus in their land. This results in periodic violent clashes with Muslims, Sikhs, and Christians.[60]

In India there has been a great deal of violence toward the hundreds of millions of untouchables or Dalits. Although the caste system has been legally banned in India, the Dalits are still looked down upon and abused. Amazingly, most of the response from the Dalits has been nonviolent, but many of them have converted to Buddhism or Christianity to escape the rejection of their fellow Hindus.[61]

As we have seen earlier, there is also a rich Hindu tradition of nonviolence. This tradition stresses that God (Brahman) is the source and sustainer of all creation, dwells in all of creation, and thus renders all reality sacred. All reality is interconnected by the divine presence, and, therefore, to damage any part of it is to hurt the self and offend against God. Union with this all-encompassing divinity is the goal of Hindu spirituality. As we have seen, the path to this union is liberation (moksha), whereby one purifies the self of all attachments to things that divide people from inner peace and union with the Divine. Another central belief in the Hindu tradition with regard to peace is ahimsa (do no harm, nonviolence). In the following sections, we will comment on these three Hindu values: the divine presence within all, liberation, and ahimsa. We will see how these values connect with efforts for peace and are implemented by Hindus who are active in peacemaking.

>> *Watch* **Mahatma Gandhi-Pilgrim of Peace**
www.gandhitopia.org/video/mahatma-gandhi-pilgrim-of
to learn further details on the life and contributions of Gandhi.

• *The Sacredness of All Beings*

The Hindu tradition teaches that the divine is the source, sustainer, and goal of all creation. The Bhagavad Gita says: "I am the beginning and the middle of all that is." This same divinity dwells within all of reality. Krishna says: "God dwells in the heart of all beings."[62] According to Krishna, everything reflects God: "Know that whatever is beautiful and good, whatever has glory and power is only a portion of my own radiance." The divine presence in all renders everything sacred and brings about a certain equality among all things: "I am the same in all beings, and my love is ever the same." To harm creation, therefore, is to offend God and hurt the self: "And when a person sees that the God that is in himself is the same God that is in all that is, he hurts himself by hurting others."[63]

• *Ahimsa as Protection from Injustice*

As we saw earlier, ahimsa (do no harm) is an important Hindu virtue and is required because of the sacredness of all things. It was Gandhi who plumbed the depth of this virtue and transformed it from a way of life largely practiced by ascetics into a force against injustice and violence to be practiced by all.

The development of a lifestyle based on a theory of nonviolence came gradually to **Mohandas Karamchand Gandhi**, commonly known as 'Mahatma' (great soul) Gandhi. When he was a lawyer working in South Africa, Gandhi was thrown out of his compartment on the train because of his color. All night he sat in the dark and cold, humiliated and now aware of what it meant to be an outcast. He decided to take up the cause of his oppressed people living in South Africa and gradually used nonviolent non-cooperation as his tool for political change. He organized strikes, protests, and marches. When confronted with violence, he trained his people to respond without violence. When he returned to India, Gandhi further developed his strategy of nonviolence and used it extensively in his efforts to gain the freedom of his country from British rule. Gandhi insisted that there be no violence: "An eye for an eye just makes the whole world blind."

Gandhi held to the Hindu tradition that all life, even the life of his enemies, came from God and was therefore sacred. He believed that God was a life force within all, and that same life force within enabled him to stand up to beatings and be jailed, humiliated, and mocked. Gandhi said: "He has never forsaken me even in my darkest hour."[64] For him, ahimsa was "one of the world's great principles which no force can wipe out."[65] Where there was nonviolence, there was Truth—and for Gandhi, Truth was God. Gandhi insisted that ahimsa not be identified with cowardice.

He believed that ahimsa was an attribute of the brave and those strong enough to stand up to abuse and even death to gain peace and justice.

Many brave people have followed his wisdom and have worked for social justice, including Martin Luther King Jr., Oscar Romero, Dorothy Day, Thomas Merton, and Daniel Berrigan.

• *Works of Holy Sacrifice*

For the Hindu, inner peace or liberation begins with the purification of self. For those who follow the ascetic life, this means rejecting pleasures, fasting, detachment from material things, and a strict regimen of yoga meditation. The Bhagavad Gita says:

> *"In success or in failure he is one; his works bind him not. He has attained liberation; he is free from all bonds, his mind has found peace in wisdom, and his work is a holy sacrifice."*[66]

For the majority of devout Hindus, liberation is sought through maintaining equanimity (shama) and nurturing the virtues of self-control, patience, forgiveness, and compassion. Inner peace must come first, and then one has the ability to help bring peace about in the world. In some schools of Hinduism, this means abandoning desires so that when acting for others there can be no discouragement, which comes from not always being successful in helping others. The very desire to succeed in helping can reveal a lack of freedom! Krishna says: "Great is the person who is free from attachments, and with a mind ruling its powers in harmony, works on the path of karma yoga, the path of consecrated action."[67] For many Hindus, the ideal is being willing to live one's life solely for the sake of others, with no attachment to praise or reward.

For Gandhi, the path toward liberation was the path toward the truth, the truth of self and of the world. For him, the search for truth was the search for God because "God is truth." It meant boring into the mystery of self and the mystery of life through action for others. This, of course, demanded a high moral standard and resistance of gratification and possession of material things.

>> **Visit the online magazine Hinduism Today:**
http://www.hinduismtoday.com/
and investigate a current topic of interest.

Hindus in Action

Mahatma Gandhi (1869–1948) – Gandhi was born in 1869 in Porbandar, India. His father was a town official and his mother was a devout Jain. He was married to Kasturbai at age thirteen and several years after decided to leave his wife and new baby to go to England to become a lawyer and English gentleman. After successfully passing the bar, Gandhi returned to his family to work in India, but was a failure as a lawyer. He heard about some law positions in South Africa and went there with his family. Once he realized how poorly the resident Indian population was being treated in South Africa, Gandhi took up their causes and spent the next twenty years defending them in court and leading nonviolent protests.

Gandhi returned to India in 1916 and spent the rest of his life struggling for freedom for the native Indian population from British rule. He was beaten, jailed, and berated, but he never turned to violence and stressed that his followers remain nonviolent. He taught his people to be self-sufficient, even to the point of making their own clothes, obtaining their own salt, and stopping the purchase of British products. He went on long fasts to stop the violence in his country. Gandhi became a national leader. Ultimately, thanks to his efforts, India gained independence. Soon thereafter, Gandhi was assassinated.

Arun Gandhi – Arun Gandhi is the grandson of Mahatma Gandhi. He was born in 1934 in South Africa where he was beaten and oppressed as a "black" by the apartheid laws. He was taught to deal with his anger against his oppressors in the ways of his grandfather: win them over nonviolently with love and courageous endurance. In particular, Arun worked with the concept of personal transformation, atma-jnana (self-realization), to learn how to work toward finding solutions to problems and avoid conflicts.

When Arun's wife, Sunanda, was not allowed to come to live with him in South Africa, he moved to India and spent thirty years working as a journalist. He and his wife spent much of their time and resources working with programs to improve the life of Indians in hundreds of villages. He now lives in the United States and speaks on many college campuses, sharing stories of his grandfather and the principles of nonviolence.

>> *Watch* **Arun Gandhi: Lessons from my Grandfather**
http://fora.tv/2008/07/18/Arun_Gandhi_Lessons_from_My_Grandfather
to learn more about self-actualization as a process for peace.

Hindu Values and Women's Issues - It often happens when religions are first established that women are given special consideration and even equality, and then in time, culture and patriarchy gradually erode these benefits. This is true in Hinduism. In the early Vedic times, women's life-giving powers identified them with nature.[68] Women wrote Vedas, participated in rituals, and had opportunities in education. Special devotion was given to goddesses.

In time, however, women's privileges were taken away. Brahmanism in northern India viewed women as impure because of menstruation and childbirth. Strict Manu laws lumped women with slaves and outcasts and required wives to be servants of their husbands. Brahmanism promoted arranged and child marriages to assure virginity before marriage, and bride dowries were required, making girls a financial liability. Women lost their legal rights for education, divorce, and inheritance. The home was to be the woman's space, and if she remained there as a faithful wife, mother, and servant, she was given certain status and authority.

From the fifth century BCE on, the Puranas prevailed, and women gained some privileges: equal access to grace and spiritual liberation independent from their roles of wife and mother. Goddesses were given a more central role in both domestic and temple rituals. In the Hindu epics there are heroines such as Sita, who at times is portrayed as independent and acts out of love toward her husband, rather than obedience. Goddesses became the role models for many women who have acted as leaders, warriors, and rebels throughout Hindu history.

Once Hinduism moved into the devotional era, women became great poets and saints. Antal (ca. 850 CE) and Mirabai (1498–1547 CE) are still held in high esteem, and their songs are sung at home and in temples. In addition to famous leaders, warriors, and poets within history, the last century has seen the appearance of "Mothers" (Mas) who have served as spiritual leaders and gurus.

Women's rights were affirmed in the post-colonial Constitution of 1948, and abuses such as child marriages and sati, where wives throw themselves on the cremation fires of their husbands, were made illegal.[69] Still, prostitution, honor killings (where women are killed for dishonoring the family), and sex trafficking are practiced in India.

Now we will look at Hindu values that pertain to women's issues and point to some outstanding Hindu women of action.

• The Divine Feminine

Many women in Christianity and Judaism would like to see the feminine included in their images of God and want to address God as Mother. The Hindu tradition has had such beliefs for thousands of years. Brahman, the ground of being who is beyond materiality and gender, is manifested in a multitude of ways, including female and male in seemingly unlimited images of gods and goddesses.

Gods and goddesses seem to reflect the infinite variety of energies within God and God's creation. Their stories and allegories tell about the human struggle and the variety of paths to salvation. If taken literally, they are fanciful and even bizarre. If interpreted correctly, they carry the wisdom of the ages.

Some Western feminists criticize the Hindu pantheon of goddesses as being ineffective in liberating Indian women, given the widespread oppression of women in India. Defenders of the goddess tradition point out that the images and their interpretations have been created by males in patriarchal cultures and can today be reconstructed and reinterpreted for today's more egalitarian culture. It is also pointed out that the goddess tradition, with all of its limitations, has for thousands of years helped women affirm their humanity, the strengths of their gender, and the sense of dignity to survive oppressive situations. The energetic productivity of Lakshmi, the consort of Vishnu, has inspired many young women in their careers. Sita, the wife of Ram, has moved many wives, including Kasturbai, the wife of Gandhi, to collaborate as partners with their husbands. Durga, who has many weapons and rides a lion or tiger as she defeats the buffalo-demon, has given many women the ferocity needed to stand up for their rights. And even Kali, who symbolizes the dark side of life and who is ferocious and bloodthirsty, has given women the strength to deal with the death of their children, to fight for the lives of their children, and to deal with pain when their children die of disease or malnourishment.[70]

• *The Growing Dignity of Women in Hinduism*
The Hindu tradition runs the gamut in describing women. One poem describes a woman as "a seed of the tree of life," and in the next verse refers to her as "a torch bearer who can lead a man to hell."[71] The Manu laws put women on a divine pedestal, but at the same time, teach that they can't be trusted. Women must be honored, but at the same time controlled and kept from bringing harm

to themselves and others.[72]

In the earliest days of the Vedic period (2500 BCE–1500 BCE), women seemed to have been an integral part of the religious services and were even authors of some of the Vedas. In these non-hierarchical agricultural communities, women were honored as symbols of fertility and nurturing. With the coming of the more hierarchical and patriarchal Aryans, women lost much of this equality. As the culture moved to urban areas with their hierarchical, political, and military structures, men gained ascendancy over women.

By 800 BCE the male Brahmins began to dominate Indian religion and society. Women lost their roles in public ceremonies and were relegated to domestic rituals, where they are still central today in Hindu homes. The impurity of women because of menstruation was emphasized, their powers of attraction were feared, and serious efforts were made to protect and control them through sequestering them at home, child marriage, arranged marriages, and even the killing of girl babies. Even though scriptures such as the Bhagavad Gita encouraged equality with teachings that offered, "I am the same to all beings, and my love is ever the same," women continued to lose ground.[73] The appearance of Buddhism and Jainism, with their more liberal views toward women, challenged the Hindu view of women, but the patriarchal values continued to prevail. It was the rise of the devotional movement, with its emphasis on passion and the heart, especially the popular devotion toward Krishna and the value of erotic passion, that sustained many women and gave them inspiration in their private spiritual lives at home.

Since independence from British colonial rule was achieved by India, women have regained some of their equality. The Constitution, as we saw earlier, is quite conscious of women's rights, and many of the former abusive practices have been outlawed. The women's movement has grown over the last century in India and is significantly attacking such abuses, especially the most serious, such as the sex-slave trade and dowry killings.

•Aspects of the Divine Mother

We have seen that in the earliest Hindu traditions the feminine was given a place of high honor. Motherhood, the source of life, was held in high esteem. Later, when procreation came to be seen as coming from the life/semen of males, with females viewed only as a depository, those honors went into decline. When nature was valued and nurtured, women became identified with it and were equally cherished (e.g., the titles Mother Nature and Mother Earth). Many think that as men began to dominate nature and abuse it, this influenced how men treated women, who had become identified with nature.

The name for goddess is devi, but the more popular name is mata or amma (mother). Mothers are considered to be "auspicious" and "beneficent," important designations in Hinduism. Many female saints and gurus have been given this title, and mothers hold a special place of honor in families.

Devi, the Mother Goddess, is worshiped throughout India, along with her many avatars or incarnations. In the Tantras she is the life force of the universe, the liberator who offers salvation to all. She is invoked in time of epidemics. She is Mother Earth and is worshiped by those who want to engage in the environmental movement. She is Mother Ganges for those who wish to go on pilgrimage, pay devotion to their sacred river, and cleanse themselves in her waters. She has been Mother India for those who wanted to liberate her from colonialists and progress as a free nation, and is Mother of the Universe for those who want to see India develop into a leading world nation. As one young Indian said as we were leaving India: "Come back in ten years. You will see how Mother has helped us build a new nation."

In modern times, a number of women have emerged who are called "Mothers," and they have attracted countless followers, both male and female. Among those are Anandamayi Ma (1896–1982), a Hindu mystic who was also a healer, and the famous Mata Amritanandamayi, known as "the hugging saint."

Hindus in Action

Anuradha Koirala – Anuradha Koirala is a former teacher from Nepal. During the past sixteen years, her group, **Maiti Nepal**, has been fighting to rescue and rehabilitate thousands of Nepal's sex-trafficking victims. As a young person she taught English in primary school. An abusive relationship led her to this crusade for young women. She says: "Every day, there was battering. And then I had three miscarriages that I think [were] from the beating. It was very difficult because I didn't know in those days where to go and report [it], who to…talk to." When the relationship ended she spent part of her teacher salary and opened a small retail shop to employ victims of sex trafficking and domestic violence. When the demand increased in the early 1990s, she started her organization to provide shelter, education, and legal defense to women. "Families are tricked all the time," said Koirala. "The trafficking of the girls is done by people who are basically known to the girls, who can lure them from the village by telling them they are getting a nice job. It's a lucrative business." Her group, many of them past victims, courageously raid brothels and patrol the India-Nepal border. Koirala and Maiti Nepal have helped rescue and rehabilitate more than twelve thousand Nepali women and girls since 1993.

>> *Learn more about* **Shabana Azmi— Women Can Change Politics** *http://www.metacafe.com/watch/2857245/shabana_azmi_women_can_ change_politics/* ***as you watch this tribute.***

>> **See Anuradha Koirala:**
http://www.friendsofmaitinepal.org/anuradha-koirala.php

Mata Amritanandamayi – Mata Amritanandamayi (b. 1953) is known as "Mother" (Amma); she has brought peace to millions. Amma dropped out of school at age nine to care for her siblings. As a child she was dedicated to taking care of the poor and elderly. She has always been a devout Hindu and has been uniquely gifted to share her love and service with others. People wait in long lines to talk to her, receive her hugs, and be blessed by her. She accepts everyone that comes to her—the diseased, the handicapped—believing that "only when human beings are able to perceive and acknowledge the self in each other can there be real peace." Amma takes in the suffering of many because she believes that "[s]orrow is the guru which takes you closer to God."

Amma's global network of charitable organizations, known as Embracing the World, has built hundreds of thousands of homes for the poor, hospitals, schools, orphanages, shelters, clinics, and soup kitchens internationally. She has also sent millions of dollars in aid to disaster areas after the 2004 Indian Ocean tsunami, Hurricane Katrina in 2005, the 2010 earthquake in Haiti, and many other areas. *Darshan – The Embrace*, a film on the life of Mata Amritanandamayi, was officially selected for showcasing at the 2005 Cannes Film Festival.

>> **Consider the video** Amma's Darshan
http://www.youtube.com/watch?v=9lwTAYeyv9U
to learn more about her healing practice and philosophy.

Listen to the words of Amma singing Everyone in the World
*http://www.youtube.com/watch?v=oANZHMov05o&feature=PlayList&p
=FF3218D44C7AE2E8&playnext=1&playnext_from=PL&index=5*
to hear Amma's hopes for the world.

SUMMARY

Hinduism, or Sanatana Dharma, has its mysterious beginnings many thousands of years ago in the Himalaya Mountains. Many scholars think that the Aryans blended their culture with that of the Indus culture and the movement produced inspired hymns, scriptures, and rituals that to this day are honored and celebrated throughout the world in diverse forms.

Hinduism, then, is a vast and complex religion. Its large collection of sacred texts includes the infallible revelations in the sacrificial songs, the Vedas and Upanishads, which explore such key notions as rebirth, action (karma), the soul (atman) and its union with the Ultimate Reality (Brahman), and human liberation (moksha). Sacred texts also include the non-infallible Laws of Manu—laws and ethics of Hinduism—as well as the great epics of *Ramayana* and *Mahabharata*. The *Mahabharata* includes the amazing Bhagavad Gita. There is also a collection called the Puranas, which develops the notion of bhakti or devotion to the gods and goddesses. The body of sacred Hindu texts also includes the Agamas, which provide Hinduism with a philosophical base for its worship and practices.

Most Hindus insist that they worship only one Ultimate Reality (Brahman) and that their numerous gods and goddesses are manifestations of this Reality. Three key gods are Brahma the Creator, Vishnu the Preserver, and Shiva the Destroyer. Many of the gods have female consorts and avatars. Some have reincarnations, such as Rama and Krishna, who are reincarnations of Vishnu. Other notable deities are Mother Durga, Hanuman the monkey-faced god, and elephant-headed Ganesha.

The ancient caste system is still operative in India and is operative in the choice for marriage and in other social situations. Although it has been outlawed to discriminate against the so-called "untouchables or outcasts," now referred to as Dalits, there is still much prejudice against them.

The Bhagavad Gita describes the three paths for liberation. The first is the path of knowledge, whereby Hindus are led by gurus in their spiritual learning, as well as in their practices of meditation and yoga. The central knowledge sought pertains to the union of the soul with the Divine and development of good karma to reach moksha or liberation from the cycle of rebirth. The second path is that of works, which include acts of puja (devotions at temple or at home), purifying the mind, renouncing sin, performing penances, and honoring the passages of life and death. The third path is that of devotion, which also focuses on rituals celebrated in temple and at home, as well as the conscientious celebration of feasts and pilgrimages to sacred places.

In recent times, Hindus have begun to reexamine their tradition in order to reclaim the beliefs and values needed to address the crucial contemporary problems of environmental devastation, violence, and the oppression of women. Today, modern India is poised to take its place as one of the future leading countries.

CHAPTER 2 VOCABULARY

Agni - The god of fire	**karma** - Action
ahimsa - Nonviolence	**mahatma** - Great soul
atman - Soul	**mantra** - Word or formula
avatar - Reincarnation of a god	**maya** - Illusion
bhakti - Devotion	**moksha** - Liberation
Brahma - The Creator	**puja** - Devotion, worship
Brahman - The Ultimate Reality	**sadhu** - Holy man, elder
Brahmin - Priestly class	**samsara** - Cycle of change
deva - Divine being	**sanatana dharma** - The Eternal law (The Indian name for Hinduism)
devi - Goddess	
dharma - Law, teaching	**sruti** - Revealed
dukkha - Suffering	**tat-etat** - This is that
ghat - Steps leading to a river	**yoga** - Yoke, spiritual practice
guru - Teacher	

TEST YOUR LEARNING

1. Explain these key Hindu terms: karma, moksha, rebirth, puja, yoga.

2. Discuss why yoga is important to Hindus and describe some of its forms.

3. Many think that Hindus worship many gods. Explain their approach to the divine.

4. Why is puja, or devotions at temple and at home, so important to Hindus?

5. What is your understanding of ahimsa, and why might it be an important belief in today's world?

APPLYING HINDUISM TO WORLD ISSUES (SHORT ESSAYS)

1. Write a short essay on how Hindu values might address some of the women's issues in India (prostitution, sex trafficking).

2. Write a short essay on how Hindu values might be applied to environmental concerns.

SUGGESTED READINGS

Berry, Thomas. Religions of India. Beverly Hills: Benzinger, 1971.

Chapple, Christopher Key, and Mary Evelyn Tucker, eds. Hinduism and Ecology. Cambridge, Massachusetts: Harvard University Press, 2000.

Flood, Gavin, ed. The Blackwell Companion to Hinduism. Malden, Massachusetts: Blackwell Publishing, Ltd., 2003.

Gosling, David L. Religion and Ecology in India and Southeast Asia. New York: Routledge, 2001.

Haker, Hille, Susan Ross, and Marie-Theres Wacker, eds. Women's Voices in World Religions. London: SMC Press, 2006.

Klostermaier, Klaus K. A Survey of Hinduism. 3rd ed. Albany: State University of New York Press, 2007.

Michaels, Axel. Hinduism: Past and Present. Princeton, NJ: Princeton University Press, 2004.

Narayanan, Vasudha. Hinduism. New York: Oxford University Press, 2004.

Nelson, Lance, ed. Purifying The Earthly Body of God. Albany: State University of New York Press, 2000.

Sharma, Arvind. Classical Hindu Thought. New York: Oxford University Press, 2000.

VIDEO RECORDINGS

Hinduism. Six videodiscs. Princeton, NJ: Films for the Humanities and Sciences, 2004.

Hinduism. One videocassette. Princeton, NJ: Films for the Humanities and Sciences, 2001.

FILMS

Briley, John. Gandhi. DVD. Directed by Richard Attenborough. Culver City, CA: Columbia Pictures, 1982.

Beaufoy, Simon. Slumdog Millionaire. DVD. Directed by Danny Boyle. Los Angeles, CA: 20th Century Fox, 2009.

NOTES

[1] Vasudha Narayanan, Hinduism *(New York: Oxford University Press, 2004), 8-9.*

[2] Thomas Berry, Religions of India *(Beverly Hills: Benzinger, 1971), 4.*

[3] *For a discussion of this controversy, see Klaus K. Klostermaier,* A Survey of Hinduism, 3rd ed. *(Albany, NY: State University of New York Press, 2007), 18ff.*

[4] *Narayanan, 38-39.*

[5] *Ralph T.H. Griffith, trans.,* Hinduism: The Rig Veda, *ed. Jaroslav Pelikan, (New York: Book of the Month Club, 1992), Book 1, Hymn 178:1-2.*

[6] *Ibid., Book 1, Hymn 63:12.*

[7] *Ibid., Hymn 5.*

[8] *Ibid., 1:113.*

[9] *Ibid.*

[10] *Ibid., 10:129.*

[11] *Ibid.*

[12] *Ibid., 10:90.*

[13] *V. P. Limaye, R. D. Vadekar, eds.,* Upanishads, Eighteen Principal Upanishads *(Poona: Vaidika Samsodhana Mandala, 1958), Mundaka Upanishad 2:1-4.*

[14] *Ibid., Katha Upanishad 4:1-5.*

[15] *Ibid., 5:6-8.*

[16] *Ibid., Brihadaranyaka Upanishad, 4:4.3-6a.*

[17] *Ibid., Isa Upanishad 4-8.*

[18] *Ibid., Taittiriya Upanishad, 2,5*

[19] *Ibid., Brihadaranyaka Upanishad 4:4.12-15.*

[20] *Ibid., 3-28.*

[21] *Ralph T. H. Griffith, trans.,* The Ramayana *(Varanasi: Chowkhamba Sanskrit Series Office, 1963), Canto 130.*

[22] *Klostermaier, 70.*

[23] *Juan Mascaro, trans.,* The Bhagavad Gita *(New York: Viking Penguin, 1962), IX: 29.*

[24] *Ibid., XII:19.*

[25] *Ibid., 2.15-20.*

[26] *Ibid., 5.20.*

[27] *Ibid., 2.56.*

[28] *Berry, 27-31.*

[29] *Natayanan, 103.*

[30] *Klostermaier, 291.*

[31] *See Axel Michaels,* Hinduism: Past and Present *(Princeton, NJ: Princeton University Press, 2004), 165ff.*

[32] *Arvind Sharma,* Classical Hindu Thought *(New York: Oxford University Press, 2000), 133ff.*

[33] *Klostermaier, 296-97.*

[34] *Flood, 497.*

[35] *Klostermaier, 165.*

[36] *Max Müller, trans.,* The Upanishads *(New York, Christian Literature Co., 1897), Part 1, Third Khanda, verse 1.*

[37] *Sharma, 92.*

[38] *Sarayana, 26.*

[39] *Muller,* The Upanishads, *IV,4,1-7.*

[40] *Sushil Mittal and Gene Thursby, eds.,* The Hindu *World (New York: Routledge, 2004), 285ff.*

[41] *Berry, 14.*

[42] *Muller, Svetasvatara Upanishad 6. 1-4.*

[43] *Mascaro,* Bhagavad Gita, *2. 39-40.*

[44] *Wendy Doniger,* The Hindus *(New York: The Penguin Press, 2009), 256-257.*

[45] *W.A. Cenkner,* A Tradition of Teacher *(Delhi: Motilal Barnarsidass, 1983), 150.*

[46] *Horace Wilson, trans.,* Vishnu Purana, *III, 12. Bing.com,*

[47] *John Renard,* Responses to 101 Questions on Hinduism *(New York: Paulist Press, 1999), 61-62.*

[48] *Jeaneane Fowler,* Hinduism *(Portland: Sussex Academic Press, 1997), 52-55.*

[49] *Ibid., 72ff.*

[50] *Christopher Key Chapple and Mary Evelyn Tucker, eds.,* Hinduism and Ecology *(Cambridge, MA: Harvard University Press, 2000), xxxvff.*

[51] *Mascaro, The* Bhagavad Gita, *7:5-6.*

[52] *Ibid., 7:7-11.*

[53] *Ibid., 10:39.*

[54] *Tucker, 5.*

[55] *Rig Veda, 10:90.*

[56] *Arvind Sharma, "Attitudes on nature in the early Upanishads," in* Purifying The Earthly Body of God, *ed. Lance Nelson (Albany: State University of New York Press, 2000), 51ff.*

[57] Bhagavad Gita, *3:25-27.*

[58] *Ibid., 13, 28.*

[59] *Christopher Key Chapple, "Toward an Indian Indigenous Environmentalism," in Nelson, 22.*

[60] *Anantanand Rambachan, "The Co-Existence of Violence and Non-violence in Hinduism,"* The Ecumenical Review 55, *no. 2 (2003): 115ff.*

[61] *Sathianathan Clarke, "Dalits Overcoming Violation and Violence,"* The Ecumenical Review 54, *no. 3 (2002): 278.*

[62] Bhagavad Gita, *10:32, 18:120.*

[63] *Ibid., 9:29*

[64] *Thomas Merton,* Gandhi and Non-Violence *(New York: New Directions, 1965), 8.*

[65] *Ibid., 32.*

[66] *Ibid., 23.*

[67] *Ibid., 7.*

[68] *Linda Joy Peach,* Women and World Religions *(Upper Saddle River, NJ: Prentice Hall, 2002), 16.*

[69] *Katherine Young, "Women and Hinduism," in* Women in Indian Religions, *ed. Arvind Sharma, (New York: Oxford University Press, 2002), 19ff.*

[70] *Fowler, 36ff.*

[71] *Madhu Khanna, "A Conversation on Two Faces of Hinduism and Their Implication for Gender Discourse" in* Women's Voices in World Religions, *eds. Hille Haker, Susan Ross, and Marie-Theres Wacker (London: SMC Press, 2006), 81.*

[72] *Werner Menski, "Hinduism," in* Ethical Issues in Six Religious Traditions, *eds. Peggy Morton and Clive Lawton (Edinburgh: Edinburgh University Press, 1996), 42.*

[73] *Mascaro, The* Bhagavad Gita, *9:29.*

CHAPTER 3

Buddhism

Ama Adhe is a devout Tibetan Buddhist. When she was a small girl, her father taught her the life and teachings of the Buddha. He took her to a lovely mountain chapel to pray and assured her that following the Buddha's path would help her survive suffering. Little did Ama know of the extreme suffering she was one day to endure. The Chinese Communists invaded her country, Tibet. They destroyed Buddhist monasteries, drove the monks and nuns into forced labor camps, and imprisoned and killed many of her people. Ama's beloved husband was poisoned; she was dragged from her two toddlers and imprisoned for twenty-seven years. She was beaten, tortured, raped, and forced into slave labor. But Ama's strong faith and dedication to prayer sustained her. Today Ama Adhe lives in exile in India where she courageously gives testimony to the oppression of her Tibetan people and bears witness to her Buddhist faith, a faith that is attracting many around the globe.

>> *Read Ama's gripping account of survival in Tibet,*
The Voice that Remembers.

Visit Ashoka—the eDharma learning center at dharmanet.org
to enroll in free courses on particular aspects of Buddhist teachings
and read recent articles on Buddhism.

Many today are drawn to the strength, peace, and compassion of the Buddhist path. At the Jesuit Catholic campus of Xavier University in Cincinnati, Ohio, Dr. David Loy, a renowned Buddhist scholar, leads his students in Buddhist meditations. One student has commented: "This hour always helps me center myself and enables me to sort out priorities and gain an inner peace."

What is it about Buddhism that attracts more than 350 million people to its practice? To answer, we need to explore the "Three Jewels" or "Refuges" of Buddhism: I take refuge in the **Buddha**; I take refuge in the **dharma** (the teachings); I take refuge in the **sangha** (the community). We will also study the divisions among Buddhists, consider the global reach of Buddhism, and observe practicing Buddhists engaging with the issues of ecology, peace, and women's issues.

THE BUDDHA

Buddha's life can be pieced together from many ancient sources and scattered biographies put together hundreds of years after his death. At the center of these legends and myths is a core biography of a real individual. Like the gospel stories about Jesus, these accounts are faith stories meant to tell the reader more about the meaning of the person of Buddha, his times, and teachings rather than historical facts about his life. The symbols used are keys to understanding the intended purpose of the story.

The Life and Myth of the Buddha - It should be noted that Buddha is not a name, or even a title. **Buddha** describes a type of person, one who has been enlightened or awakened to the truth. While many Buddhists believe that through the eons there were and are many buddhas, one stands out as the Buddha. In history, he was a real person named Siddhartha Gotama. Gotama was born in approximately 448 BCE, in the village of Lumbini, located in the foothills of the Himalayas near the border between India and present-day Nepal. He lived for approximately eighty years, although scholarship varies on his exact age. His father was a leader in the Sakya, a warrior kingdom, so Buddha is often called Sakyamuni, the Sage of the Sakya.

The stories of Siddhartha's conception and birth are magical. At his conception, his mother was transported in a dream to the Himalayas, where a magnificent white elephant carrying a lotus flower entered her side. She was surrounded by flowers and streams of water from the heavens, great bursts of light, and the sounds and sensations of an earthquake. An early text states that a hermit declared that this child was an extraordinary person and predicted that he would reach the heights of enlightenment in the world and would spread his compassion and teaching to many.[1] Early Buddhist sources say that a number of Brahmins predicted that the child would either stay at home and become an emperor or become a wandering ascetic. The extraordinary birth occurred as Gotama's mother stood under a tree in a lovely grove filled with birdsong. The infant then stood up and spoke of his noble destiny.[2]

>> *Google* **The Life of Buddha** *and watch the video to gain further insight into the Buddha's path to enlightenment.*
The Life of Buddha, 50 min - Jul 13, 2007 *http://www.youtube.com/watch?v=YsEksMEE2Eg*

• *Born a Wealthy Prince*

Siddhartha was the son of a king, a prince who enjoyed an opulent and sensuous life. He was a member of the royal and wealthy class in a society that was rigidly divided between rich and poor. The wealthy kingdoms were swallowing up local villages and forming large urban areas where the rich lived in abundance, while the poor existed amid hunger, squalor, and disease.

One legend says that Siddhartha's father, having heard the prophecy that his son would either be a powerful leader or a monk, decided power and wealth would be his son's destiny. To achieve this, the father did everything he could to prevent his son from becoming a reclusive monk. He gave Gotama three palaces and surrounded him with luxuries and beautiful women. Gotama was given a lovely wife, Yasohara, who bore him a son, Rahula. They were the ideal family, comfortably isolated from the harsh reality outside the palace.

• Encountering Reality

Tradition tells us that Siddhartha took three forays in his char-
iot to see the real world. In spite of his father's efforts to clear
the areas of distasteful sights, "the gods" saw to it that Gotama
came into contact with an old man, a sick man, and a corpse. Siddhartha saw the dark side of life and realized that his own youth,
health, and idyllic family life were fleeting, impermanent. None
of the things he had could offer him ultimate happiness; it would
all end someday. Gotama became disillusioned with his ideal life
in the palace. He was drawn to become one of the renunciates,
men of his time who lived an ascetic life.[3]

Siddhartha became dissatisfied not only with his home life, but
with his religion as well. When Siddhartha was born, the dominant religion was Brahmanism, which would eventually become
the religion now called Hinduism. This religion seems to have
been brought to the area by the Aryans perhaps one thousand
years earlier. It was a religion of many gods, dominated by priests
(**Brahmins**) who presided over magic rituals and sacrificial fires,
designed to manipulate the gods to keep order in the world and
provide health, wealth, and immortality. Many, especially the
downtrodden, often felt that the gods were unresponsive to their
desperate needs.

The Brahmins, who were usually identified with the wealthy
and powerful, placed themselves at the top of the order of castes.
The order then descended to warriors, then to cattle-herders and
farmers, and finally to the bottom order assigned to servants and
slaves. Status came from one's birth, and belief in reincarnation
indicated that **karma**, or one's accumulated merit from a past life,
decided the level into which one would be born. Generally, the
lower castes viewed themselves as caught in an endless cycle of
undesirable reincarnations, such as being reborn as an animal
or an insect. Their only hope was to become associated with the

magic and sacrifices of the Brahmin priests, but, as the lowest caste, the poor had little access to the magical fires and ceremonies. Few were taught to have faith in their own inner resources, and most felt helpless in their dependence on gods who were apparently unable to help them.

As we shall see, many of the Brahmin beliefs influenced Gotama, but his interpretations differed. He would set aside the gods, goddesses, and the rituals and turn to reason and self-effort for salvation. He would replace the Brahmin belief in an immortal soul with the acknowledgment of a consciousness that could live on. Buddha would replace the Brahmin fatalistic notion of endless cycles of reincarnation with the deliberate and chosen rebirth of the consciousness. Buddha rejected the Brahmin caste system. His notion of karma was not based on strict adherence to religious orthodoxy as was that of the Brahmins; he was more concerned with freedom from suffering and the purification of the consciousness. At the same time, Gotama would stress the Brahmin importance of asceticism and intense meditation for self-purification and nonviolence (ahimsa).[4]

• Leaving It All

Gotama decided to leave his family and join the disillusioned living as monks in the forests or as wandering beggars searching for the truth. His parents were brokenhearted to see Gotama put on the saffron robes of a monk. He slipped out at night without even saying goodbye to his wife and child. Buddhists refer to this as "The Great Going Forth."

First, Gotama sought out yoga teachers in Magadha and Kosala so that he could begin to master the ancient art of meditation to gain insight. Gotama began an intense six-year period of mortification. He practiced long non-breathing activities until his head and stomach ached with pain. He endured long fasts until he became gaunt and weak and could no longer think straight. Buddha commented that his backbone stood out, his ribs were visible, his eyes were sunken deep in their sockets, and the skin on his head was shriveled. Gotama eventually discovered that such extremes did not help him in his search, so he accepted rice and mild gruel from a kind young woman, gradually restored his health and energy, and moved on.

• *The Enlightenment*

The tradition tells us that Gotama one day came upon a Bodhi tree in Bodh Gaya and decided to meditate there until he found the truth. During his meditation, he passed through many stages of concentration. One story in the tradition points out that Gotama underwent severe tests from Mara, a mighty evil spirit, who tried to tempt Gotama with power and sensual desires and even challenged Gotama's very right to be under the tree. Like the Christian story of Jesus' rejection of Satan in the desert, Gotama resisted Mara's temptations to give up his search, go back to the merit system of Brahmanism, and take up a life that was sensual, fearful, and lustful. Gotama called upon the earth to assist him (Buddha is often portrayed motioning toward the earth), and Mara fled.

Throughout the night Gotama reviewed his past lives and the future of all living things. At dawn, Gotama arose as the "Buddha," the enlightened one. He had awakened to overcome ignorance and craving. He had moved beyond the suffering attached to old age, sickness, and death; he had achieved the ultimate state of peaceful nirvana. As Thomas Berry put it:

> *"This moment in the life of Buddha is one of the most significant moments in the spiritual life of mankind. The illumination that broke over his mind has ever since flowered over the greater part of Asia in a great tide of spiritual healing. For centuries the people of Asia have walked in the radiance of this light."* [5]

The enlightenment, the "waking up," or nirvana is the ultimate goal in Buddhism. It is a deep realization that all things are impermanent, and that this awareness is the ultimate cause of human suffering. Gotama saw that we become attached to people, places, and things, and as we inevitably experience the loss of everything we cherish, we suffer anguish and pain. Buddha woke up to the insight that the only way out of such suffering is to "let go" or give up our attachments, and be compassionate and kind. The noble life, according to the Buddha and his followers, would be one that is centered on the other, rather than on the self and one's needs. It called for a radical transformation in the way one experiences one's self and the world.

Buddha remained in Gaya for several more days, savoring the joy of his enlightenment. Then he faced the next challenge: Was he willing to spend the rest of his life teaching others what he had learned? Would it not be easier to simply live in the forest as a recluse rather than to weary himself trying to teach people who could not understand?

Buddha now knew that ignorance was *the* obstacle that caused humans to wander endlessly in the dark. Would people be so mired in ignorance and so given over to their desires that they could not accept his difficult teaching? Should he launch on this difficult mission to help others become a Buddha and awaken to the truth of life?

• *The Mission*

Buddha decided that he was called to spread his teaching (**dharma**) and would dedicate himself to calling those who had faith in Brahma, the god of creation, the "Deathless One," to set faith aside and focus on self-transformation.[6] His mission would be "for the welfare of the manyfolk, for the happiness of the many-folk, out of compassion for the world, for the good, the welfare, the happiness of gods and men." [7]

The tradition tells us that Buddha went to the Deer Park (deer are still there today) in the holy Indian city of Sarnath and sought out the five recluses with whom he had lived. There he taught them the Four Noble Truths and the Noble Eightfold Path. This format seems to be an easy-to-remember shorthand for the core teachings of the Buddha, assembled and synthesized in a manner similar to the way the Christian gospel writers put together Jesus' teachings from the Sermon on the Mount.

Buddha's message was as welcome as clear running waters after a long drought. His teaching was not doctrine, but consisted of aids to understanding how people can be liberated from suffering. Buddha would not provide a set of beliefs or a manual of rules for worship, but a path and a guidebook. Social barriers would be torn down and the caste system rejected; everyone would be welcomed to listen and join in—the nobles and the servants, the rich and the poor, the established and the outcasts. Kings, Brahmins, Jains, forest monks, ascetics: all were welcome. True nobility was to be achieved not by birth or possession but by spiritual growth. Buddha said: "Birth neither Brahmin, nor non Brahmin makes; 'tis life and conduct moulds the Brahmin true."' [8] Or, as Martin Luther King Jr. put it: "judged not by the color of their skin, but by the content of their character."

The dharma was good news, especially to the oppressed who could now achieve their salvation through their own efforts. There would be no need for empty rituals, sacrifices, magic, or heavenly grace from the gods who seemed to be helpless. People could develop their own inner resources and freely participate in the search for nirvana, or ultimate happiness. No longer would there be a need to be afraid of endless cycles of reincarnations. Inner peace and happiness (nirvana) could be achieved in the here and now by meditating, living a moral life, and being ever mindful of the real truth that underlies the illusions of reality.

Buddha traveled extensively, teaching and gaining new followers wherever he went. He established orders (**sangha**) of ordained monks and nuns and organized lay followers. He was known for his humility, wisdom, and compassion and had unique psychic and healing powers. At the same time, not everyone approved of Buddha's influence. Many resented his popularity and his large following. Some objected to how he was calling men away from their families to the ascetic communities. There were several attempts on Buddha's life, but he was always miraculously saved.

In his last days, Buddha suffered from a serious illness, but kept pressing on. Accidental food poisoning seems to have been the ultimate cause of his death, and he died lying peacefully on his side under two trees. Before his death, Buddha asked that his corpse be wrapped in a cloth and cremated, and that the ashes be placed in a stupa, or burial mound, where people could come to experience confidence and happiness. Stupas have now become shrines for the Buddhists. His final words were: "All individuals pass away. Seek your liberation with diligence."

After the cremation, Buddha's ashes were gathered in bowls and distributed to eight large shrine-like stupas throughout India. The Buddha did not pick a successor, wanting the dharma, or teaching, to be what future generations would follow. Yet Buddha would remain forever to his followers an extraordinary man and even, for many, a divine presence.

Some Meanings in the Myth — Stories of Buddha's life are teaching stories, and many of the details bring the heart of Buddhism to light. The miraculous happenings that surround his conception and birth (the flowers, streams, deities) symbolize both the natural and cosmic significance of this birth. One thinks of the Christian story of the star, Magi, and angels singing at Jesus' birth symbolizing the wonder of "God's Son" coming into the world.

The wealth and sensuousness of palace life seem to symbolize the Buddhist belief that clinging to such material and physical "illusions" of happiness only brings suffering. The young prince's going forth to seek reality teaches the need to emerge from sheltered living to search for the truth. At first viewing, Gotama's abandonment of his wife and child appears heartless and irresponsible. But the action is symbolic of the need to detach from the comforts of home and family in order to be free enough to establish independence and personal identity. Buddhist monks must leave all and live as homeless wanderers in search of inner enlightenment. The fact that a woman helped bring him back to health seems to symbolize the important role women would play in the community.

Gotama's initial practice and then rejection of extreme asceticism represents the Middle Way that is the Buddhist path. It is a path midway between luxury and rigidity. It is the path of moderation, where one's basic needs are taken care of, but which avoids both the trap of luxury and the degradation of squalor.

The meditation of Gotama under the Bodhi tree has many layers of meaning. The notion of "tree" is highly symbolic in other scriptures. In one Hebrew creation story, there is the "tree of life" and the "tree of good and evil." The first was available for food; the second contained the forbidden fruit that would bring death into the world. In the Christian passion stories, Jesus is crucified on a tree (cross) and through that death brings salvation. In the Buddhist tradition, at Buddha's birth, enlightenment, and death, the tree is a symbol of new life and new light.

In the stories, Buddha's enlightenment does not come from the gods or from supernatural power. It comes from his natural powers to go within himself and find the truth and light of reality. This becomes central to the Buddhist way. It is the intense effort through regular meditation to purify and set aside the self in order to discover inner peace and happiness.

The stories show that even the Buddha was hesitant to carry out his mission of overcoming stubborn ignorance. Yet he overcame his reluctance and became a role model for every follower on the daunting mission of bringing kindness and compassion into a world that is often violent and merciless. Buddha's ability to deflect attacks on his life seems to symbolize the indestructibility of this movement.

Finally, even Buddha had to suffer death. Everything and everyone is impermanent. The material world and all human egos attached to it are mere illusions. The Buddha passes on—and yet lives on in his presence and in his teachings. He can lead the way along the path toward happiness.

Originally the Buddha was considered to be a man, an extraordinary man, who became enlightened and brought to the world a final and conclusive revelation on how to overcome suffering and reach enlightenment. Several centuries after Buddha's death, some of his followers began to view the Buddha as a glorified transcendent being, still present in the world and continuing his ongoing revelation. Buddhahood was now in everyone, and all were called to strive to be a Buddha.[9]

THE DHARMA

Buddha began to teach in what some call the Axial period (800–200 BCE), a pivotal time (described in chapter 2) when new ideas burst forth; a time when major philosophical and religious teachings were expounded by important figures. Besides the appearance of Buddha, this was the time of the great Hebrew prophets of Israel, Mahavira of the Jains, the Upanishadic sages in India, Confucius and Lao Tzu in China, Zoroaster in Persia, and Socrates and Plato in Greece. This indeed seems to have been the beginning of a new era.[10]

The Four Noble Truths – The Four Noble Truths are a summary of Buddha's teachings (dharma) on the universality of suffering, the causes of suffering, and how people can be freed from suffering and achieve deep peace and happiness. These truths can be summarized as:
1. Suffering is everywhere.
2. Suffering is caused by desire.

3. Suffering can cease by eliminating desires.

4. Desires can cease by following the eightfold path of right living.[11]

• *The First Noble Truth*

The first Noble Truth of Buddhism is concerned about the truth of suffering (**dukkha**). Buddha taught that suffering is integral to human life: birth, old age, illness, death, grief from the loss of loved ones, the presence of things we hate, and not being able to get what we want.

Everyone is familiar with some form of suffering. Parents hear that their child has a brain tumor; a young mother is diagnosed with breast cancer; someone fails an exam, loses a job, or has a breakup with a lover. Parents hear that their son was killed in battle. Dukkha refers to the dark side of life: the bitterness, pain, and sorrows of life. It also refers to a deeper awareness that nothing is permanent. Ultimately, we will lose everything and everyone that we love. As Buddha put it: "That which begins, also comes to an end."[12]

This realization of impermanence in itself can bring us anxiety, like the existential dread people experienced after World War II. Forty million people had died, millions had been exterminated in concentration camps, and much of Europe and Asia lay in ruins. Reactions to this caused a deep angst that was expressed in the writings of philosophers such as Sartre and Camus. Many were moved to say that life was absurd, or as Shakespeare put it: "a tale told by an idiot." Such an emphasis on suffering does not signal that Buddhism is negative or pessimistic. Buddha taught that we were meant to be happy and joyful, but because all things are impermanent, we are subject to loss and suffering.

Change and flux are integral to life; any attempts to avoid them are futile. Individuals can say: "Carpe diem" (seize the day), or "Eat, drink, and be merry because tomorrow we die," or (simply falling into self-pity) "Woe is me!" According to Buddha, none of these escapes will bring people to the happiness that they seek. He believed that he offered a true path to peace of mind and happiness, the way of "strict austerity, a holy life, the Noble Truths seen clear—Nirvana won."[13]

Liberation from suffering is at the heart of Buddha's teaching. The Buddha once commented:

"Brethren, the mighty ocean has but one flavor, the flavor of salt. Even so Brethren, has my teaching one flavor, the flavor of release [from suffering]." He summed up his mission as follows: "Suffering I teach—and the way out of suffering."[14]

Buddha's life mission, then, was to bring about the cessation of suffering, but not through denial or escape. The tradition says that we should not push our suffering aside, but rather hold it in our arms like we would an infant. Then we should look at the suffering tenderly and ask: "What is causing my suffering? And how can I set aside these causes so that I can be happy again?"

For more than 2,500 years, billions of disciples have followed Buddha's path and have found healing for their suffering in goodness, loving kindness, and compassion.

This is not to say that Buddha thought he could eliminate suffering. Rather, Buddha's goal was to teach people how to free themselves from the grip of suffering through mindfulness, where we live within the present moment in peace. Buddha taught that humans can be architects of their own consciousness. Through meditation and unselfish living, they can root out toxic thoughts and emotions and dedicate their lives to unselfish service of others. The Buddhist way is one of truth, goodness, and inner peace. The earliest tradition said that nothing could compare with discovering "this gem the Buddha holds; and may this truth bless all! The perfect Peace he preached, our Sage's deathless peace, can find no match elsewhere."[15]

• Differing views on suffering

Throughout the ages there have been various interpretations of the Buddhist view of suffering. For some, "everything is suffering," and they view suffering as the basic experience of life. For some Buddhists this has resulted in flight from the world into forests and monasteries. For others, suffering is only part of the human experience. The well-known Buddhist monk Thich Nhat Hanh suggests that the slogan "everything is suffering," often attributed to Buddha, is not in fact his teaching. He adds that while there is a universality of impermanence, that is not the case with suffering. He observes that while suffering does come into our lives, it is not always a part of our experience.[16]

Another Buddhist approach to suffering is that it can be redemptive. In this approach, suffering can be endured for others and can be the very instrument of their liberation. Take, for example, the case of Aung San Suu Kyi, the courageous Buddhist woman who won the Nobel Peace Prize in 1991 for her struggle against the brutally repressive regime in Myanmar (formally Burma). Dr. Suu Kyi has been a staunch advocate for human rights in her country and has vehemently protested against the beating, imprisonment, and killing of her fellow Buddhists. Her positions resulted in her being blocked from becoming prime minister and being held under house arrest for most of the last twenty years. She was released in 2010 and remains a courageous symbol of liberation for her people today.

• The Second Noble Truth

The Second Noble Truth is concerned with causes of suffering. Buddha taught that desire or craving is the first cause of suffering. We know that everything is impermanent, but we can act as though this is not true. Desires or cravings drive people to hold on to things as though they are permanent. Buddha puts it bluntly: "The creatures of desire, mere slaves of mundane joys, shall scarce be saved."[17] Buddha humorously compares this craving to apes bounding through the trees: "Chopping and changing fails to calm distracted folks, who now hold fast to this, and now to that, like apes that skip from bough to bough."[18] Besides craving, there is a human drive to cling to people and things, "to hold on for dear life," knowing full well that we really cannot hold on forever. So often, deep suffering is attached to trying to cling to a relationship after a breakup. Or pain comes to a parent who will not let go of a son or daughter who has left home. Buddha teaches that this craving and clinging is a main source of anguish, and his mission was to teach people how to heal such unhappiness.

But many Buddhist teachers believe that desire in general is not the cause of suffering. Thich Nhat Hanh points out that evil desires attached to anger, ignorance, arrogance, greed, pride, and misperceptions can bring us to grief. One thinks of the enormous amount of suffering caused by the Nazi hatred of the Jews, by the bitterness of Communist regimes in the Soviet Union and China toward religious people, or by the greed of some on Wall Street and in the global financial markets before the economic collapse of 2008.

• *The Third Noble Truth*
The Third Noble Truth regards being liberated from suffering. This involves giving up desires and evil cravings, as well as avoiding clinging to people and things. It means letting go of the illusion of "forever" or "happily ever after," which moves individuals to cling to people, things, the past, or even life itself. While loss cannot be prevented, the perception of loss can be altered. As priest and eco-theologian Thomas Berry said:

> *"The problem of sorrow is to be solved, not by altering the external conditions of life, but by strengthening people inwardly, enabling them to rise above the world of change to an inner*

experience that would permanently remove them from sub-jection to the agonizing experience of temporal existence."[19]

This is the path that rejects the illusion of permanency and turns within, to the self, for true inner happiness.

• *The Fourth Noble Truth*

The Fourth Noble Truth points to the path toward liberation from the causes of suffering. The path is eightfold, often symbolized by the eight frames or ribs on the umbrella that monks carry to shade themselves. The path is also symbolized by the Buddhist Wheel with its eight spokes. This is the path of rightness in all aspects of life, the path of avoiding evil and doing good, the path to nirvana. All eight elements of *rightness* (views, thought, speech, action, livelihood, effort, mindfulness, and concentration) are to be followed simultaneously in life and not as steps in a moral plan.

Right views, thought, and speech pertain to **wisdom**, or having a proper view of reality, its impermanence, suffering, and liberation. Here a deeper understanding of the truth about life is the goal. The right view sees that all things are impermanent and cannot be held on to as lasting without causing suffering. The right view perceives that ignorance, hatred, anger, and other evil feelings can bring untold sufferings to the self as well as to others. Right thoughts and right speech follow from this view and ensure that toxic thoughts are not entertained and that harmful speech is not used.

Right action and livelihood pertain to morality. Buddhist morality calls for proper ethical choices and actions flowing from these choices. The Buddhist basic precepts are to avoid killing living beings, stealing, sexual misconduct, lying, and the use of intoxicants. Right livelihood requires that one select a profession that is carried out with honesty and integrity and without harm to others. In short, Buddhism recommends a healthy lifestyle, where one lives a good life and tries not to harm the self or others.

Right effort, mindfulness, and concentration are concerned with **meditation**, the constant practice to purify the self and to reach deep within where truth exists. This requires that the disciple makes regular efforts to find a quiet, secluded place to meditate. It calls for a disciplined mindfulness, whereby the disciple can also meditate while eating, walking, or working. Peaceful mindfulness of the present moment is key in Buddhism for acquiring inner tranquility. Concentration refers to the ongoing focus toward well-being and service of others. In the Dhammapada, Buddha teaches that meditation brings us to the virtuous life, which has more fragrance "than of sandle wood (sic), rosebay, of the blue lotus or jasmine."[20] And, of course, it is virtue or the lack of it that determines one's karma.

Karma – Buddha teaches that, although nirvana can be achieved in this life, most people require a series of rebirths to achieve this state. Here, Buddhism is influenced by the Brahmin notion of the **law of karma**, which determines the nature of these rebirths.

Karma refers to one's actions. The ancient belief in the Indian tradition held that causes produce effects; therefore, there are consequences to all actions. Good actions result in good results and in shaping good people. Bad actions bring about bad results and shape bad people. In biblical terms, we might say: "You reap what you sow." In the Western world, we might say: "What goes around comes around."

In both the Hindu and Buddhist traditions the notion of karma became integrated with the belief in rebirth. Many in the Hindu tradition believe that souls are reborn countless times, and that whether or not the soul advances in the quality of rebirth depends on the nature and quality of the soul's actions. Acts of violence, for instance, could result in a rebirth in a series of hells or places of purification, and the rebirth might be in the form of an animal. Acts of greed might result in rebirth as a ghost. By way of contrast, acts of kindness and compassion bring about a more favorable rebirth, perhaps as a good person in a virtuous family.

Buddha put it this way: "No man's deeds are blotted out; each deed comes home; the doer finds it waiting for him in worlds to come."[21]

There is a difference between the Hindu and Buddhist approach to karma. For the Hindu, it is the soul that acts and accumulates karma. The Buddhist tradition, on the other hand, does not posit the existence of a soul. For Buddha, it is the consciousness that acts; therefore, the intention is more important than the action. In ethical thinking this was a revolutionary breakthrough—the motivation or intention of an action became more important than simply the external action that was done.

Meditation – In Bodh Gaya, India, at the Buddhist Temple, it is striking to watch a Buddhist nun in her tailored gray outfit, with a shaven head and serene face, taking measured steps in a walking meditation. The slow movement of her slim body, the peacefulness of her countenance, and the careful placement of each foot concretely demonstrate true mindfulness.

Regular meditation is at the heart of Buddhist practice. While the Western world is occupied with exploring outer space, the Buddhists are dedicated to exploring "inner space." It is a way of purifying the mind, the inner consciousness. For the Buddhist, it is this consciousness, not a soul as it is for the Hindu, that can be subject to rebirth.

Meditation aims to bring about evenness of thinking and feeling, peacefulness and joy within, and the ability to live deeply in the present moment. It usually involves a regular practice of calmly sitting or walking, where one puts aside distractions by focusing on breathing. The past is gone, the future is unknown; concentration centers on the present moment. What are our concerns, and are they healthy or morbid? What are our feelings, and are they toxic to our well-being or positive and life-giving? What are we suffering, and what are the causes of this suffering? Is my hatred of someone or my desire for revenge causing me anguish?

How can I cast these destructive desires aside for my own peace of mind? What are the sufferings of those around me, and how can I bring them relief through compassion and kindness?

When concentration and mindfulness are attained through meditation, wisdom is achieved. That wisdom guides our views, thoughts, and choices. Regular meditation throughout the day is the Buddhist way to remain centered, thinking and acting with a perspective that is good, compassionate, and loving. As Maha Ghosananda, the heroic monk from the times of the killing fields in Cambodia, put it:

> ... the cause and condition of violence is greediness, anger and ignorance. The cause and condition of peace is non-greediness, non-anger and non-ignorance....To overcome ignorance, we have to practice wisdom.[22]

For the Buddhist, meditation is the way to mindfulness, the means to live deeply in the present moment. The past is gone, the future is unknown. The present moment is all that exists, and it needs to be purified of toxic thoughts and feelings, so that the inner joy and peacefulness that lies deep within each person can be discovered. Meditation is the "gathering in" of all that leads us to peace, and the "cutting out" of all that hinders us from peace.[23] Going forth with such a frame of mind and heart enables one to be a source of goodness and kindness to others.

>> **Listen to the following chant and experience living in the moment.**
See Om Mani Padme Hum on YouTube:
http://www.youtube.com/watch?v=bk6q0zxa4xQ&feature=.

The Discourse on Loving Kindness, one of the great spiritual treasures of Buddhist literature, puts it this way:

> *"Just as a mother with her own life protects her son, her only son, from hurt, so within your own self foster a limitless concern for every living creature. Display a heart of boundless love for all the world....Devote your mind entirely to this; it is known as living here a life divine."*[24]

The No-Self – Another Buddhist goal is achieving the **no-self**. The Buddhist notion of the no-self is extremely difficult for Westerners to understand, since they put so much value on personal individuality and independence.

For Buddha, people are subject to many illusions, including the illusion of the self. For Buddha, the so-called *self* is nothing more than a fluid and dynamic process, ever changing. It is a construct of matter, feelings, perceptions, and thoughts that are constantly in flux. For example, from the Buddhist perspective, Jocelyn the little girl who jumped rope and played with her dolls is a different person than the young Jocelyn who goes to college, or the Jocelyn who gets married and has children, or the Jocelyn who launches a career as a counselor, or who in old age sits quietly alone in a retirement home. Here Jocelyn is considered to be a process, rather than a person, and once stripped of her roles and masks is a *no-self*, a selfless person who, once stripped of the *I am*, can be free to be given to others in compassion and love. Buddha says: "Let him pluck out obsession's root—the craze: 'I am.'"[25]

Buddha teaches that the self is an illusion. It is not the elements of the body, because these die and decay. Nor is it what we see or hear, because sights and sounds fade away. Neither is the self the consciousness, because this stops when we are asleep. It is not the present mind, because that lasts but a moment. Nor is the self the future mind, because that does not yet exist.[26] For Buddha, the goal is to discover the no-self. Perhaps the closest a Westerner can get to this notion is "the true self," the inner self, stripped of all masks, roles, and pretenses. Or we might refer to the selfless self, the self given entirely for others.

Nirvana – The ultimate goal in Buddhist practice is to achieve nirvana. This state seems to be impossible to explain, just as mystics can never clearly describe their encounters with the Holy. The Buddha told his followers that it was foolish to speculate about the nature of nirvana or how to achieve it. He likened such a questioner to a person shot with a poisoned arrow, who would not let the doctor cure him until he knew every-thing about who shot the arrow and what it was made of. While looking for such answers, the man died![27]

The Buddha reported that once he himself had reached enlightenment he knew that after death he would enter nirvana—free from all suffer-ing, free from the cycle of rebirth, beyond all perception and all attach-ment. He told them that his final birth had already come about and that he would not have to endure any future bodily life. Buddha also told his disciples that nirvana was a condition where there is no materiality, no space, no consciousness or perception. He said: "It is not in this world, nor in another world…there is neither coming nor going, no staying or leaving, no being born or dying, no stillness or movement. It is the ces-sation of all suffering. There is no self, no falsehood….In this condition there exists only tranquility."[28]

Buddha described nirvana as the emptiness, wherein people experi-enced a clearing away of all illusions, including the illusion of ego. It is liberation from the desires and limitations that cause so much suffering in life. It is a perception of the "pure heart, refined and straight, free from obstacles, free from low cravings, that the disciple of the Noble Ones has fully attained…."[29] The ancient sources remark that nirvana is "deep, im-measurable, unfathomable, like the great ocean."[30] It is beyond love, intel-ligence, faith, or understanding.

Nirvana is, in fact, beyond self and being and becoming. It is the deep experience of the true reality that lies beneath the illusions that appear to be reality. It is the end of the cycle of rebirth. To describe nirvana, the tradition uses the symbol of a candle flame going out, marking the quiet end of crav-ing and suffering. In the words of the Buddha: "As flame blown out by the wind, even so is the sage."[31] Christians might hear echoes here of the perfect freedom and happiness that they call "heaven," or the "nada" (nothingness) that the Christian mystics see as their goal in contemplation.

The Dharma Is Also Impermanent – The Buddha pointed out that both he and his teaching were also impermanent. He constantly stressed that all teaching should be tried and verified through experience, not on the word of some external authority or even the word of the Buddha. He encouraged his disciples to think independently: "You are to be lights unto yourselves. You are to be refuges unto yourselves....Do not look toward anyone but yourself for refuge."[32] The Buddha believed that all teaching had to be tested by critical thinking, and that is one reason that he remains a great educator today.

The Buddha also taught that teachings should be of practical value, and that once the goal was achieved, teaching could be set aside. He compared his teachings to a raft used to get to the other shore. Once the crossing is achieved, there is no need to carry the raft on one's shoulders. It can be left behind. So it is with teaching.

Buddha was always open to disagreement with his teaching, yet he was quick to correct someone who misinterpreted or distorted his message. At the same time, he was wise enough to know that throughout the ages his message would be interpreted many different ways as times and circumstances changed. He knew that his movement would have to adapt to new cultures and circumstances. Buddha's disciples should be "unprejudiced and free, not based on learning's stores, owning no sect or school, holding no theories."[33]

No God Talk – God and the gods are at times mentioned in the early Buddhist tradition. It is important to note that Buddhism is not atheistic. Buddha opposed the inadequate God-talk of his time, but he did not deny the existence of God. Buddha seems to have believed that theological talk of divinity and eternity was an irresolvable distraction. His goal was the extinction of suffering, not to develop a theology.

While Buddha seldom used the word "God," he did embrace the Ultimate, the Absolute, and yet developed a completely different way of expressing the ultimate from that of his Brahmin predecessors.

THE SANGHA

The **Sangha**, in some traditions, refers to the monks who live an ascetic life of celibacy, simplicity, and prayer. The monks are ones who are expected to study the teachings of the Buddha and be able to instruct others. They are to be the role models for the laity.

The monks live simply, with few possessions, and are committed to a life of fasting, celibacy, and abstinence from alcohol. In some traditions, one can become a monk as a boy. The commitment need not be for life, and many men live as monks for just part of their lives. The lay people respect them for their detachment and put food in their begging bowls in order to gain merit. At festivals, the laity bring the monks cloth for new robes, food, and other necessities of life.

In some traditions, the sangha has a broader meaning, referring to the community of monks, nuns, and lay people. Here, all are an example to each other and see reflected in each other the peace, joy, kindness, and compassion of the Buddha. Taking refuge in such a sangha, or community, brings one a sense of security, trust, and belonging. Like any religious community, the sangha provides social, psychological, and spiritual support. It provides concern in times of trial and support in times of need. Buddha always told his followers to look after each other with kindness and compassion.

As we shall see, the later divisions in Buddhism involved questions about the nature of the sangha. Were the monks that lived the life of Buddhist perfection central in the religion? Were the laity who existed as outsiders only able to gain by serving the monks? Or were both laity and monks seeking perfection in their own ways? Was the raft (the vehicle used to cross to enlightenment) only for monks (*Hinayana*–lesser vehicle, a pejorative description of Theravada Buddhism), or was it also for the use of the laity (*Mahayana*–greater vehicle)?

Soon after Buddha's death, his monks called a council to preserve and continue his teachings. At that council several of Buddha's closest associates recited the oral traditions of his teaching in order to establish an authentic set of teachings. As subsequent councils were called, disputes arose regarding these teachings. The first signs of differing schools of thought appeared. There was serious controversy about the rules of the monasteries, and a schism occurred. Buddhism evolved and adapted to new cultures. Many interpretations emerged, and the tradition became quite diverse and divided. And yet, with all of its variations, there seems to be a universal core at the heart of all forms of Buddhism—how to let go of all desires and avoid suffering.

The original split (100 CE) was largely concerned about whether the emphasis in Buddhism should be on the rigid monk's way of life or the more populist, devotional way of the laity. Both movements legitimately pointed to teachings of the Buddha to justify their positions. While Buddha himself emphasized the monastic life, he also opened his teaching to all people and encouraged self-reliance and personal effort in the search for enlightenment.

During his lifetime, Buddha was able to keep both of these perspectives in balance, but after his death each view began to have a life of its own. Theravada and Mahayana Buddhism became the two major divisions. Within each of these, a number of schools of thought evolved. Buddhism now exists with many diverse and complex traditions.

Theravada Buddhism – In the same year of Buddha's death (approximately 500 BCE), the Council of Rajagriha was called to discuss Buddhist doctrine and monastic discipline. At this Council, the arhats, or monks who had achieved spiritual distinction, dominated and emphasized the importance of the monk's life within the Buddhist tradition. It was stressed that to enter enlightenment, one had to leave home and live as a recluse according to monastic discipline. This position left the laity in the secondary position of living an inferior devotional life, wherein they were expected to be both obedient to and supportive of the monks. At a second Council held one hundred years later, this same emphasis on the primacy of the monastic life was reiterated.

Soon after the second Council, the laity reacted and a schism began to build. Arhats began to be criticized for their sexual impurity, ignorance, and doubts. During the time of the Emperor Ashoka (250 BCE), who favored the laity and intended to spread that version of Buddhism to his kingdom, the divisions grew more noticeable.

The conservative Theravada movement hardened its position, resisting any liberal views that Buddha was a celestial being. Theravada stressed that the Buddha was a man—a unique and exceptional man—but he was still only a man, whose personal influence came to an end when he died. Theravada Buddhists do not pray to the Buddha, but rather they offer homage to him as the great teacher and example of the Buddhist way.

As for practice, the followers of the Theravada movement stressed the necessity of the homeless state, celibacy, abstention from alcohol, and living a strict monastic life. Here, the arhat, the perfected disciple of Buddha, was expected to renounce the world and live a simple life in a monastery, unencumbered by material things. This was to be a life of intense meditation, regular concentration and mindfulness, and a constant search for nirvana.

Theravada, also referred to as "The Way of the Elders," bases its position on the early Pali Canon, a collection of writings of Buddhism. These writings stress the fact that the Buddha left his home and family and established an order of men who led a disciplined life, emphasizing that being a Buddhist is a full-time commitment and must be carried out without the attachment, distractions, and pleasure of the lay life. Theravada stresses independence of the individual, wherein nirvana can be achieved solely on one's own through discipline, and without the help of anyone and any higher power.[34] The Buddhist road to the no-self is, therefore, free from what others can make us into and from roles that others expect us to play. The human is away from the world, entirely cut off from worldly concerns, material gains, or human expectations. Wisdom is the key virtue to be sought, and it is best achieved by denying the self and purifying the consciousness as one strives for enlightenment.

Theravada holds the laity in a secondary position. They are relegated to a devotional life, and are expected to follow a basic, but inferior, moral code. The laity gather in temples and shrines, where they bow before the image of Buddha, light candles, hold up burning sticks of incense, place flowers, and offer prayers. The laity are expected to support the monks. Normally, the laity come to the monasteries to celebrate the great Buddhist feasts with the monks, bringing food and clothing for the community. For these gifts, the laity can expect to receive merit for their karma. And when the monks circulate the neighborhood with their begging bowls, the laity share food with the monks and thus hope to gain additional merit. From the monks, the laity can expect spiritual instruction, education, good example, and officiating at key events such as births, marriages, and deaths.

Theravada is referred to as Southern Buddhism because it spread to Sri Lanka, Myanmar (formerly Burma), Thailand, Laos, and Cambodia. In most of these countries, Theravada is once again growing since Communist suppression has subsided in this region of the world. Theravada has also spread to the West, where its followers do not usually live in a monastery, but gather regularly in a temple for meditation and retreats.

Mahayana Buddhism – The Mahayana school of Buddhism arose from an earlier tradition called Mahasanghika, which exalted the Buddha as a celestial being who came to the earth as a savior. Mahasanghika also shifted the centrality of Buddhism from the monks to the laity and was far more open to miraculous events and the power of devotion.[35] This tradition, which eventually developed into the Mahayana, claimed to be rooted in the original tradition equally with Theravada Buddhism. At the same time, the Mahayana school was open to the authenticity of later documents and comfortable with new interpretations. This led to the acceptance of an ongoing and developing notion of Buddhist tradition. Once this dynamic and more populist tradition appeared as the Mahayana, around the time of Jesus (30 CE), it became widely appealing and eventually spread to Nepal, Tibet, China, Korea, Mongolia, Vietnam,

Japan, and later to countries in the West.

Many Mahayana Buddhists are devoted to the historical Buddha, but also experience him as a divine presence. At the same time, they often acknowledge that the Buddha-nature exists in any person as a seed and can be nurtured and brought to fruition. In the Mahayana tradition, mindfulness is getting in touch with the Buddha that is within.

Thich Nhat Hanh, a leading Vietnamese monk, writes:

"We need to touch the Buddha within us. We need to enter our own heart, which means to enter the heart of the Buddha. To enter the heart of the Buddha means to be present to ourselves, our suffering, our joys and for many others."[36]

This mindfulness can be achieved in walking, talking, meditating, and living everyday life. It involves being present to the self, others, and the Buddha in compassion and love. These latter virtues are central to Mahayana, rather than the Theravada, wisdom.

In Mahayana, the bodhisattva, the new title for disciples, replaced the more traditional arhat. The bodhisattvas vigorously spread their teachings and sought out new converts. Lay people gradually began to play a more prominent role in Buddhism and could be bodhisattvas. The bodhisattva enters the path to become a Buddha, but is willing to postpone his or her own liberation in order to liberate others through teaching and good deeds. Whereas Theravada stresses independence, Mahayana stresses dependence on others and on the power of divine grace. Salvation is not for the few, but for the many, and people are saved or liberated through the compassion of others. The emphasis here is not in fleeing the world, but in being part of it and in participating in its liberation. Here, one becomes the no-self by giving the self to others, rather than withdrawing from the world in order to find the no-self within. And emptiness, rather than being a goal as it is in Theravada, is a means of clearing out all the obstacles that prevent the disciple from being a compassionate and loving person.

Both Theravada and Mahayana are authentic traditions of Buddhism, but the emphasis that they place on elements of the core beliefs differs. The renowned leader of Tibetan Buddhism, the Dalai Lama, has clarified how Theravada and Mahayana are linked together and rooted in the same tradition.

He writes:

All of Buddha's teachings can be expressed in two sentences. The first is "You must help others." This includes all of the Great Vehicle (Mahayana) teachings. "If not, you should not harm others." This is the whole teaching of the Low Vehicle (Hinayana). It expresses the basis of all ethics, which is to cease harming others. Both Teachings are based on the thought of love and compassion.[37]

Vajrayana Buddhism – A number of Buddhist traditions developed from Mahayana. One is the **Vajrayana**, or the Diamond Vehicle, which has come to be known as Tibetan Buddhism. When Buddhism was brought to Tibet in the seventh century CE, it was blended with the shamanism of the area, and a unique tradition evolved, which used prayer flags, elaborate rituals with masks, the ringing of bells, and the turning of prayer wheels. It also embraces mystical practices concerning the dead.

Eventually, potent Tantric methods (use of the divine power or prana that flows through the universe and one's body to attain purposeful goals) were brought from India and taught by expert gurus called lamas. New methods were developed to reach entrance into ultimate reality and to harness the unconscious forces within the person. Some thought that these energies could be used to reach enlightenment within one's own lifetime. Vajrayana Buddhism practice follows a three-stage path of meditation, compassion and wisdom, and the diamond vehicle. It is called the diamond tradition because it aims at developing mind into a light-filled reality that is indestructible. The Dalai Lama is the most well-known Tibetan Buddhist.

Pure Land Buddhism – Pure Land is one of the great Chinese schools of Buddhism, although it is also widely practiced in Japan. Pure Land Buddhism developed the devotional life of many in Eastern Asia and deeply influenced the practice of Buddhism. It finds its source in a bodhisattva who became an enlightened one called the Amitabha Buddha (the Amida Buddha in Japan). This Buddha, once he attained enlightenment, created a Pure Land where people can be saved from rebirth and come to awakening. Attainment of this bliss is possible through devotion to Amitabha Buddha.

Pure Land was founded by a patriarch called T'an-luan (476 CE–542 CE), who learned about it from a Buddhist monk. T'an-luan took on the practice and began Buddhist teachings based on the Land of Bliss Sutras that had been translated into Chinese around 230 CE. These teachings showed disciples how to picture and call upon the Amitabha Buddha to achieve rebirth into the Pure Land. This devotion had popular appeal because it gave the common folk, even evil people who wanted to repent, access to rebirth. Just calling on the Amitabha Buddha could purify and bring light into one's life. Calling on the name of the Other could release a person from being self-centered and enable a person to help others enter the Pure Land.[38]

Shan-tao was another great teacher in this tradition. He further developed this tradition to include the notion of empathy toward the human suffering brought on by evil actions. Shan-tao taught that if one has true repentance and humility, one can turn to the Amitabha in faith and be reborn into the Pure Land. This piety had a broad appeal because the ordinary person could call on Buddha in faith and be led into the infinite light of the Pure Land, not only after death, but in this life as well. Jodo (Pure Land) and Jodo Shinshu (True School of the Pure Land) are schools that developed in Japan in the thirteenth century CE.

Nichiren Buddhism – Nichiren Buddhism was founded by a young man born in a small fishing village in Japan in 1222 CE. He lived near a Shinto shrine and began to wonder why the gods had not saved the emperor from a recent uprising. In his search, he tried various Buddhist traditions and finally decided that the Tendai school based on the Lotus Sutra was the best. This is a sutra thought to have been delivered at the end of Buddha's life. Nichiren became convinced that if people became liberated through the teachings of the Lotus Sutra, there would be peace and harmony in his country. His negative attacks on other Buddhist traditions made him enemies, and he was exiled several times. Once pardoned, he started a group of followers who would gather to chant the opening of the Lotus Sutra: "I take refuge in the Lotus of the Wonderful Law Sutra." They believed that the chant had the power to put them in touch with their Buddha-nature and lead them to the goal of a good moral life and Buddhahood.[39] The chant is believed to connect followers with the divine and is repeated many times a day. They also were convinced that following this practice had the power to bring peace to Japan and transform it into an enlightened and harmonious nation.[40] In modern times, Nichiren Buddhists have been extremely active in the peace movement in Japan.

Zen Buddhism – Zen is a Japanese word for meditation, which today is one of the main practices of Zen Buddhism. Zen Buddhism seems to have originated with Bodhidharma, a monk from India, who brought this tradition to China in the sixth century CE. There it blended in with the native Taoism. Later, it spread to Japan, where it has had a profound influence on that country. Zen is a branch of the Mahayana tradition, and while it tends to avoid texts and words so that one can be moved by experience, this school is devoted to the teachings of the Buddha. Its Zen Masters are well-versed in the Buddhist Sutras.

One focus of Zen Buddhism is that each person has the Buddha-nature and wisdom within, but contact with it is blocked by delusions and anxieties. In other words, when we think things are going wrong, they aren't really wrong at all. It is just that we are not in touch with the inner peace, wholeness, and wisdom that have been within us from the beginning. For Zen, we are not seeking perfection, but trying to get in touch with the perfection that is already there. This connection is how one achieves enlightenment.

There are various ways Zen attempts to get in touch with inner wisdom and move toward enlightenment (satori in Japanese). One way is zazen, or sitting meditation. Quietly one sits, listens to one's breathing, centers the self, and reflects on such questions as: Who am I? Why am I the way I am?

The meditator works to let go of thought patterns that lead away from the truth or that separate the meditator from honesty.[41] The focus is to be in the present and allow oneself to flow in the living moment of the now. Effort is also made to set aside dualities such as outside and inside, good and bad, so that we can experience a unity within as well as around us. In the Rinzai sect of Zen, koans, or puzzling and paradoxical statements, are used to interrupt the logic of thinking and free the person to think along new and creative lines. Here is an example of a koan: "One day as Manjusri stood outside the gate, the Buddha called to him: 'Manjusri, why do you not enter?' He replied, 'I do not see myself as outside. Why enter?'" Suddenly the one meditating is stopped and must think about what is truly outside and inside.[42]

Other means are used by Zen Buddhism to accomplish the formidable task of quieting the mind and opening the consciousness to its innate wisdom. At times, the Zen Master shouts or hits the desk with a stick to interrupt distractions. Lovely pen drawings of nature, well-cultivated gardens, tea ceremonies, and chanting are all used to quiet the mind and free it to go inward to the true self and a genuine sense of reality. At times, the art includes comic portrayals of gibbon monkeys, pictures of old straw hats and shoes, or beautiful flowers—items that will stop the mind and the heart and set them on a new and refreshing path.

Zen Buddhism was made popular in the United States by D.T. Suzuki (1870–1966). His works and teachings had a profound effect on Americans, and as a result there are many Zen Centers throughout the country. These centers offer instruction in Buddhism, opportunities for meditation led by a Zen Master, and retreats for more intensive experience.

Buddhism Spreads – The most effective proselytizer of Buddhism was the Indian Emperor Ashoka the Great, who ruled from 273 BCE–232 BCE). Lamenting over the slaughter perpetrated during his military conquests, Ashoka was drawn to the compassion and nonviolence of Buddhism. His edicts became a model for just, tolerant, and nonviolent governments. During his reign Buddhism spread throughout his vast empire.

For centuries, Hinduism and Buddhism flourished side by side in India. By the eighth century, Buddhism began to give way to Hinduism and Jainism in western and southern India. Buddhism's decline accelerated in the tenth century with new developments in Hinduism and with Islamic invasions, which did not take lightly to what was considered to be "idolatry."[43] Buddhists fled to south India, Nepal, Tibet, and Sri Lanka. Buddhism spread to other countries early on. A century before the Christian era, it expanded along the Silk Road to Central Asia. It appeared in Sri Lanka around 250 BCE and has flourished there in spite of oppression by colonialists. In the early centuries of the first millennium, Buddhism spread to China where it linked with Confucianism and Taoism and emerged as Pure Land Buddhism; it then spread to Cambodia, Vietnam, Korea, Japan, and Thailand.

>> *Visit the Buddhist Channel* http://www.buddhistchannel.tv *for current news on Buddhism.*

Visit this website to view Story of India:
http://www.youtube.com/watch?v=SihssFt8XGs

Watch the introduction to the film Cry of the Snow Lion *at*
http://www.dailymotion.com/video/x7op9g_tibet-cry-of-the-snow-lion_music
for a taste of Tibet's struggles and triumphs.

MODERN MOVEMENTS IN BUDDHISM

In the twentieth century, Buddhism experienced serious persecution and decline. Buddhism has suffered greatly at the hands of Communist governments. With the Communist takeover in China in 1948, many temples and monasteries were destroyed and their lands confiscated. During Mao Tse-tung's Cultural Revolution (1966–1972) in China, Buddhism was targeted for annihilation. Since 1977, the Party has been more tolerant; some monasteries and temples have been allowed to open, but they can only function if registered with and controlled by the government.

The Chinese invaded Tibet in 1950, claiming it as a Chinese territory. In 1959 the young Dalai Lama fled to India, where he now lives in exile in Dharamsala. Since 1966, six thousand Tibetan monasteries have been destroyed and more than 1 million Tibetan Buddhists have died, either during rebellions or from famine and persecution.[44] The Chinese are still making serious efforts to destroy the Tibetan culture, largely by sending large numbers of Han Chinese, the primary ethnic group in the People's Republic of China, to live in Tibet and by forbidding Tibetan education.

In Vietnam, Buddhism lost ground during the war (1959–1975), but its renewal has been tolerated under the present Communist regime as long as there is registration with the government. Periodically, there are incidents of repression. In October of 2009, 150 followers of Thich Nhat Hanh were driven from a monastery near Hanoi because they gathered at a memorial for Vietnam War dead.

Maha Ghosananda (1929–2007) was a highly revered Cambodian Buddhist monk. Highly educated, Ghosananda spoke fifteen languages. He was one of the few monks to survive Pol Pot's slaughter of Cambodians during the Vietnam War. Ghosananda heroically ministered to his people, and after the war helped to restore Buddhism and establish peace in his country. Maha Ghosananda has been called the Ghandi of Cambodia.

>> **Watch the video detailing the communist takeover of Cambodia, Return to the Killing Fields:** http://www.youtube.com/watch?v=sRpGJ2pPjlM

In Cambodia, most of the temples were destroyed and many of the monks were killed during the terror of Pol Pot (1975–1979), the infamous Communist leader of the Khmer Rouge. He ordered the murder of more than 1 million citizens to "purify" his culture. Today, there is a revival of Buddhism in Cambodia. Some temples are being rebuilt, and the number of monks is growing, although many of them are poorly trained and not qualified to be leaders in their communities. Little is known of Buddhism's fate in Communist North Korea.

In the non-Communist areas of Taiwan, Singapore, and South Korea, Buddhism is flourishing. In Japan, the so-called Buddhist New Religions are gaining ground, especially the Soka Gakkai branch of the Nichiren sect. Buddhism is growing in Thailand and in Sri Lanka, especially since the twenty-year-long war with the Tamils has ended. Buddhism is the dominant religion in Myanmar (formerly Burma). Recently, there have been clashes between the restrictive military regime and the Buddhist monks about the spiking of fuel prices, which especially impacted the poor. Some monks were beaten and imprisoned, while others went into exile.

There is a growing popularity of Buddhism in the West. Many temples and communities are flourishing in the United States, Canada, England, Wales, Scotland, Northern Ireland, and throughout Western Europe. This was given impetus in the 1960s by the Beat Generation and also the Beatles, who were attracted to Buddhism. Today, its popularity has attracted celebrities such as Richard Gere and Harrison Ford, who follow Buddhism and champion the cause of Tibet. We will discuss Buddhism in the West at more length in the next section on social issues.

CONNECTING BUDDHISM TO WORLD ISSUES

We have looked at the ancient Buddhist tradition, its founder, basic teachings, divisions, and expansion in modern times. In China, Buddhists face the challenges of reviving a tradition that was nearly destroyed by Communism and confronting a growing consumerism and militarism that opposes their values of peace and simplicity. In Japan, Buddhists face a strong secularism and materialism that challenges their commitment to spiritual practices and meditation.

Many Buddhists throughout the world have decided that practicing their faith through chanting, meditation, and ascetic living alone is not enough. Contemporary Buddhists are applying their beliefs to contemporary problems such as terrorism, poverty, discrimination, sexual promiscuity, and the sex trade. Tibetan Buddhists are deeply distressed by the Chinese take-over of their country, as well as China's efforts to destroy their ancient culture, where Buddhism has always played a prominent role.

Here we will discuss in more detail how Buddhists worldwide are becoming more involved in three issues: the environment, peace, and women's issues.

Buddhist Values and Ecology – Buddhists today are increasingly concerned about their environment. In a number of Asian countries, they have formulated five- or ten-year plans for improving their mountains, forests, and cities. In Mongolia, where for seventy years the Communists destroyed most of the monasteries and their lands, new monasteries are being built, and the land is being restored. Trees and flowers are being planted, and the monks are being trained to teach their neighbors not to hunt the endangered snow leopard and saiga antelope, to avoid pollution, and to recycle. As we shall see in more detail later, monks in Cambodia and Thailand, as well as Tibetan monks in exile in India, have become powerful activists in the environmental movement.

Buddhists are making many connections between their beliefs and their responsibilities toward the earth. At the same time, many Buddhist leaders are cautious. Though they realize that Buddhist teachings have often adapted to new cultures and problems, they do not want Buddhist dharma to be distorted just to follow trendy "green" or "new age" movements. They do not want the emphasis of interconnectedness and service to neglect the Buddhist value of self-formation. Many Buddhist leaders do realize, however, that their traditions are living, dynamic traditions and can be responsibly interpreted in new ways to address current issues.

In the following section, we will look at some of the Buddhist values that are being linked to ecology.

• *Interdependence*

Especially in the Mahayana tradition of Buddhism, the law of causality and interdependence is emphasized. The Buddha was noted for his teaching that cause and effect is the basis for the constant change we observe in life. Every action causes a lasting reaction, and this is the basis of karma. Our thoughts and actions make up who we are. In addition, all cosmological and natural systems operate on the principle of cause and effect. All things in the cosmos are intimately linked together as one whole. Although there is a hierarchy of beings in Buddhism, superiority and domination are rejected—all must work in cooperation. The disciple is to have "an all embracing for all the universe for all its heights and depth..."[45] The ancient Mahayana Buddhist metaphor of Indra's Jewels (also called Indra's Net) conveys the notion of interdependence. It describes a many-jeweled net or web in the heavens. As light reflects on the facets of these jewels, each element in the universe is reflected in the many facets, and each jewel reflects the whole cosmos. The stars, sun, moon, rivers, trees, animals, people, indeed all living creatures, have their own reflections and shine on each other. Light here is used as a magnificent metaphor for causal relationships. In Buddhism, "the enlightened one" teaches that all sentient beings share in the same fundamental conditions of birth, suffering, old age, and death. This universal condition of suffering produces compassionate empathy, which liberates a person from suffering. It is a call to see life holistically and to live in harmony with nature.

>> **Visit Ecobuddhism.org for information and slideshows on a Buddhist response to global warming.**

This belief in interconnectedness is beginning to be applied to ecology by many Buddhists. As one environmentalist put it: "Buddhists have sometimes taught us to be fearful of nature. They have suggested that the world is an endless net of causality where every event sends ripples throughout the whole fabric of the universe. This may be a healthy lesson, demonstrating that we need be more fearful of the consequences of what we do with regard to nature."[46]

• *Mindfulness*

Another valuable truth that Buddha offers those concerned about the earth is the need for mindfulness on the delicate balance in nature—the vulnerability of the earth and all that lives on it. Buddhism calls people to learn that the "enemy" is not outside of us, but within us. Buddhism urges people to restructure their consciousness and root out their fears, their toxic thoughts and practices, their feelings of apathy and helplessness. The tradition invites people to be mindful of the careless and dangerous path they trod and calls them toward loving kindness and compassion for their earth and its people. Its ancient tradition of "doing no harm" is especially relevant here. The Dalai Lama has observed that we humans are the only species with the power to destroy the earth as we know it.

Buddhists in Action

Sulak Sivaraksa – Sulak Sivaraksa was born in 1933 and is a prominent Buddhist monk from Thailand. He is an intellectual and a social critic, and has written many books including *Conflict, Culture, Change: Engaged Buddhism in a Globalizing World* and *The Wisdom of Sustainability: Buddhist Economics for the 21st Century*. Sulak is very critical of Western capitalism and the destruction it is bringing to the environment. He maintains that the poor are especially affected by the degradation of the earth.

Sulak has founded a number of organizations that work for indigenous, sustainable, and spiritual models of change. He has been arrested and exiled several times for his criticism of the Thai government. During these times, he has taught at American and Canadian universities. In 1995 Sulak won the Right Livelihood Award, known as the Alternative Nobel Peace Prize.

Sulak teaches that Buddhism is a questioning process, questioning everything including the self. His mission is to offer Buddhist principles and practices as a personal and political resource for world change.

>> **See Sulak Sivaraksa**: *http://www.youtube.com/watch?v=OklerY9LelE*

Jiyul Sunim – Jiyul Sunim is a South Korean Buddhist nun who is a well-known environmental activist. She turns her compassion toward nature. Jiyul has gone on life-threatening fasts and organized demonstrations to challenge the building of a tunnel for high-speed trains, which she claimed threatened the ecosystems of Mount Cheonseong in South Korea. Her Buddhist values have helped the people of South Korea become aware of how development can destroy their environment, and she has been a strong advocate for the rights of nature. Jiyul believes the Buddhist teaching that as one lights a fire, or listens to leaves shaking in the wind or to a mother calling her child, one suddenly understands the relationship between oneself and all the beings in the world, and one is awakened. She points out that even though all things are impermanent, they are the key means through which we achieve enlightenment and are worthy of our compassion and care. She believes that enlightenment can be found in the sound of the wind and the waters. Most recently she has fought development along the Nakdong River and has also been an advocate for eco-feminism. She started the **Green Resonance Movement** in South Korea for protecting the environment.[47]

Buddhist Values and Peace – The Buddha was committed to a mission to end suffering and bring peace through compassionate and kind living.

In one of the most treasured Buddhist collections, the Dhammapada, the Buddha says:

> *Better than a thousand hollow words is one word that brings peace. Better than a thousand hollow verses is one word that brings peace....It is better to conquer yourself than to win a thousand battles. Then the victory is yours.*[48]

None of the earliest Pali texts support violence, and there are only several later texts that allow violence to protect life, though these are highly disputed.

In practice, there have been many occasions where Buddhists have turned to violence. Throughout the centuries, the Buddhists in Sri Lanka have been at war with the Hindu Tamils, and only recently came to a peaceful resolution. Japan had its "warrior monks" and shogun fighters in the Middle Ages. In the modern era, most Buddhist schools supported Japanese nationalism and Japanese involvement in World War II. It was Zen Buddhism that developed the martial arts for fighting. Some Buddhists turned to violence against their Communist enemies in Cambodia, Vietnam, Thailand, and Myanmar (formerly Burma). For a half century the Tibetans have struggled to resist the occupation of their country. Even now, the Dalai Lama struggles to discourage his young followers from turning to violence against their Chinese occupiers.

In spite of these examples of Buddhist violence, the overwhelming consensus among the scholars of Buddhism is that Buddhism is against violence.

• *Nonviolence*

The Buddhist path is one where anger, hatred, or greed (the causes of violence) are unacceptable to the human psyche. Killing, plainly and simply, goes against the first Buddhist precept of "do not kill."

From Buddha's perspective, peace begins with peace of mind and then spreads to others through teaching and example. The Buddha identifies the nonviolent person as the holy one: "He who has renounced violence toward all living things, weak or strong, who neither kills nor causes to kill—him do I call a holy man."[49]

The Buddhist path to harmony and nonviolence calls for rooting out the causes of violence that exist within each person. Ignorance is the first cause of violence because it entails clinging to false notions of reality, notions based on an illusionary individuality, a centered ego, and a disordered love of self.

Anger, especially in reaction to abuse and hatred, is viewed as another driving force for violence. The Buddha taught "hatred is never stopped by hatred." There are stories of Buddha himself being abused, and how he suffered patiently and without retaliation. At one point his own people were about to enter a war over water rights, and he intervened, pointing out that human lives are more valuable than material things. His alternatives to violence were patience and compassion. He taught: "As a bush fire burning out of control stops only when it reaches a vast body of water, so the rage of one who vows vengeance cannot be quelled except by the waters of compassion."[50] Ultimately, it is greed that cuts us off from oneness with others and leads to violence and suffering. The Lotus Sutra says: "The cause of all suffering is rooted in greed. If greed is extinguished, there will be no place for suffering."

• *Orientation toward peace*
One branch of Tibetan Buddhism calls for a three-fold personal orientation toward peace: renunciation, transformation, and self-liberation.

Renunciation occurs when bitterness, self-pity, grudges, desires for revenge that drive persons to destroy others, and other toxic thoughts are rooted out of the consciousness. The Dalai

Lama calls this internal disarmament. He writes: "When, as individuals, we disarm ourselves internally—through countering our negative thoughts and emotions and cultivating positive qualities—we create the conditions for external disarmament."[51]

Transformation embraces the virtues of no harm (ahimsa), loving kindness toward all, and the realization of the interconnectedness of all things.[52]

Self-liberation allows practitioners to step away from the absolutes and dualisms that cut them off from other living things, from the good/evil, enemy/friend, I/it, or even I/thou, rather than the "I am you" reality. The suffering and pain of others becomes one's own, and one reaches out in compassion. The goal here is achieving the "no-self."[53] This does not mean the extinction of the self, but rather the transformation and liberation of the self. The "no" here is the rejection of the greedy and violent self and the embrace of the Buddha-self that is kind and compassionate. As the Buddha teaches: "when one's desires are quelled, their passing ushers in the calm of peace."[54] Only then is the true disciple willing to sacrifice the self to save others from suffering.

The Mahayana tradition stresses that the peaceful person is able to generate peaceful energy to the small community (sangha) and then that energy can move out from there to the larger community.[55] Here one becomes a bodhisattva and is willing to postpone nirvana for the self in order to serve others. As mentioned earlier, the disciple gets in touch with her or his Buddha-nature, acknowledges the Buddha-nature in all living beings, and relates to them accordingly. This, of course, precludes harming or killing other living beings.

Many leading Buddhists today have strongly opposed violence and war. They are committed to the Buddhist value of ahimsa (nonviolence) and to the teachings of the Buddha: "Victory breeds hatred; the defeated live in pain; the peaceful live happily, giving up victory and defeat."[56]

Buddhists in Action

Dalai Lama – The Dalai Lama is the leader of Tibetan Buddhism and is a renowned figure throughout the world. For more than fifty years, he has been leading his followers from his place of exile in Dharamsala, India, in nonviolent ways to gain justice for his Tibetan people. In 1998 he won the Nobel Peace Prize for his tireless efforts for peace. His comments on war were candid:

"Chairman Mao once said that political power comes from the barrel of a gun. Of course it is true that violence can achieve certain short-term objectives, but it cannot obtain long-lasting ends. If we look at history, we find that in time, humanity's love of peace, justice, and freedom always triumphs over cruelty and oppression. This is why I am such a fervent believer in nonviolence. Violence begets violence. And violence means only one thing: suffering."[57]

| **See Dalai Lama talks:** http://www.youtube.com/watch?v=20MnLcOL7Ks

Maha Ghosananda – Maha Ghosananda (1929–2007) was a Buddhist monk and a much-revered hero in Cambodia. He valiantly stood with his people during the horrors of the Killing Fields under Pol Pot's reign of the Khmer Rouge (1975–1979).

Educated in Cambodia and India, he received his Ph.D. at Nalanda University. In the 1970s the Khmer Rouge regime in Cambodia imprisoned, tortured, and killed most of the Buddhist monks. Ghosananda left his forest monastery and heroically stood with his people in exile camps in Thailand.

When the terrors of the Khmer Rouge ended, there were very few monks or monasteries left in Cambodia, and Ghosananda's family, friends, and fellow monks had been wiped out. He stepped up and worked tirelessly to rebuild Buddhism and his country. He led peace marches and pilgrimages to restore the Buddhist tradition and became a shining light for his people. He has been called "the Gandhi of Cambodia," and was nominated three times for the Nobel Peace Prize.

Thich Nhat Hanh – Thich Nhat Hanh is a Buddhist monk who participated in the peace movement during the Vietnam War. He courageously led a Third Way movement that refused to take sides in the conflict and insisted on nonviolence as a means to end the conflict. As a Buddhist, he refused to accept the dualism of win/lose or friend/enemy. He applied the Buddhist notion of karma to demonstrate the evil results that came from violent actions. It was his conviction that everyone is a victim in war, and that peace and human life must be put before everything. Hanh is now a world-renowned Buddhist writer and teacher. He continues to teach peace from Plum Village, a Buddhist monastery in France. His motto is: "Peace work means, first of all, being peace."

Chan Khong –

| Read and report on Chan's gripping book, *Learning True Love: How I Learned and Practiced Social Change in Vietnam.*

Watch Chan Khong discuss her work in Vietnam:
http://www.youtube.com/watch?v=df0gWhCrb0g

Chan Khong is a Buddhist nun who worked closely with Hanh during the ravages of the Vietnam War. As a young woman, she helped orphans and refugees during the struggle between the French and the Vietnamese Communists. Later, she was persecuted by both the Catholic regime as well as the Communists in Vietnam. She now works closely with Hanh in Plum Village in France, where she carries out her work for hungry and homeless children.

Aung San Suu Kyi, who was blocked from taking office as prime minister of Myanmar (Burma) in 1990, spent many years in prison and under house arrest, separated from her husband and two children. She is pragmatic about the use of nonviolence and has chosen it because it protects her people from further harm. Suu Kyi won the Nobel Peace Prize in 1991 for her courageous struggle for peace.

Buddhist Values and the Women's Movement – The 300 million Buddhist women in the world are on the move. After one thousand years of refusal, they have in some places regained the right for nuns to be fully ordained. They have established Buddhist universities and schools and have recruited millions of young nuns and laywomen to learn and practice the Buddhist tradition. They have built orphanages, hospices, retirement homes, elementary and high schools; they have organizations to improve the environment, work for peace, fight domestic violence, end sex trafficking, and provide animal rescue.

• *The Buddha's Honor for Women*

The Buddha, of course, stands as the standard for enlightenment for his followers, so his views on women are of great interest. In the Buddhist tradition, his mother, Maya, is honored, as well as the woman who gave him food when he was starving. Buddha taught that the path to awakening, or liberating the consciousness from suffering, could be followed by both men and women.[58] This belief that women are capable of enlightenment has been intrinsic to Buddhism since the beginning. The Buddha taught that the mind had no intrinsic gender and that, therefore, women were just as capable of achieving enlightenment as men. In places such as China and Vietnam, this is symbolized by the highly revered Quan Yin, the Buddhist goddess of compassion and mercy.

Even though the Buddha taught that women were capable of enlightenment, it took him some time before he would agree to allow women into the Order. It took the efforts of Pajapati, the aunt who raised him after the death of his mother, to convince him. Buddha turned down her request three times, driving her to take drastic measures. Pajapati gathered a large crowd of women, who shaved their heads, donned the saffron robes of monks, and marched for one hundred miles to Buddha's home to present their request for ordination. When the Buddha's close disciple Ananda saw the women exhausted and covered with dust, he appealed to the Buddha. Buddha finally agreed that the women could be ordained, with stipulations: the women must follow eight additional rules and be subordinate to the monks.[59] Even with these provisos, Buddha's decision and his overall view toward women went against the extremely patriarchal positions of his day. This has given contemporary Buddhist women encouragement to strive for equality.

Early on, many nuns (bhikkhunis) established themselves as teachers, spiritual masters, poets, and leaders.[60] One of the most famous was Sanghamitra, the daughter of Ashoka, the great Indian emperor who spread Buddhism throughout Asia. Along with sixteen other nuns, Sanghamitra traveled to Sri Lanka and firmly established Buddhism in that country. The nuns lived in many parts of Asia until the Middle Ages, when they disappeared from the scene.

During the twentieth century, Buddhist nuns began to reappear, first in Nepal, then in Vietnam, Taiwan, Tibet, India, China, and throughout Asia. There are now, also, a significant number of Buddhist nuns in the United States and Canada. Many of these nuns, especially those in Asia, live in very poor circumstances with little access to donations and with few opportunities for education in the Buddhist tradition and in the professional skills necessary for serving as teachers, nurses, or religious leaders. Needless to say, there are often fewer opportunities for Buddhist laywomen. Still, in countries such as Taiwan, South Korea, Vietnam, and Singapore, the opportunities for women, both lay and monastic, are becoming more available.

• *Belief in Liberation*
In the modern era, Buddhist women are becoming more empowered by the teachings of the Buddha. His goal "to liberate all sentient beings" has moved many Buddhist women toward liberation from the oppression, inequality, abuse, and discrimination of women that exist in their cultures. Most Buddhist women are Asians and resist calling themselves feminists because they associate the term with white, middle-class Westerners. Like black women who often refer to themselves as "Womanists," or Latinas who often use the term "Mujeristas," Asian women see their challenges and goals as different from those of feminists. And even among Asian women, the obstacles to equality can differ from country to country.

Buddhist women have to think long and hard about the Buddhist teaching of no-self since many have had enough of being self-effacing and humble servants to the men in their communities. They can, however, accept the valid struggle to be unselfish and to use their abilities to alleviate the suffering of others, as long as their personal human dignity and freedom are clearly recognized.

Buddhist teachings move women toward a new vision of liberation in the modern era. Following the tradition, they struggle to be free of anger and hatred toward the men who oppress them, knowing that such feelings are toxic and have to be rooted out in disciplined meditation. Buddha's teaching of the interconnected moves Buddhist women to seek closer relationships with their Buddhist sisters. Their sense of oneness frees Buddhist women to serve the less fortunate in their communities.

• *Equality and Independence*

The Buddha taught his followers to "be a light unto yourself," which gives women a sense of independence and encourages them to be critical thinkers. He also taught that women were equal to men. The reassertion of these values today has given Buddhist women a new sense of dignity and liberation. Whether they are bending in the rice paddies from dawn to dusk, rowing tourists in the Mekong Delta, working in factories in Ho Chi Minh city, or studying or teaching in a university, Buddhist women today are gaining a new sense of themselves as being free, equal, and independent. Buddhist women worldwide are in an era of tremendous revival. Throughout Asia, and in Europe and North America, more women are becoming devout laywomen or choosing the monastic life. As Buddhist women gain greater access to education in the Buddhist tradition, enter professions, and join the workforce, they will indeed become influential in bringing Buddhist teachings to education, health care, peace activities, social work, and ecological projects.

Buddhists in Action

Karma Lekshe Tsomo – Karma Lekshe Tsomo founded **Sakyadhita (Daughters of the Buddha)**, an organization that provides many Buddhist women with education, inspiration, and support. Since the founding of Sakyadhita, the organization has met every other year in Asian cities. At each of the last two meetings in Vietnam and Singapore, more than 2,500 Buddhist women met to present scholarly papers, network and offer support, meditate, and learn. A strong sisterhood, the organization has become a major force in promoting the causes of Buddhist women all over the world.

>> **See Karma Lekshe Tsomo and view her addresses :**
www.youtube.com/watch?v=GbAijRlg0-E

Khunying Kanitha Wichiencharoen – Khunying Kanitha Wichiencharoen (1920–2002) is a Buddhist nun who tirelessly served abused women and children in Thailand, a country known for prostitution and sex trafficking. Wichiencharoen, an accomplished lawyer, established the well-known **Emergency Home for Women in Distress** in Bangkok, which has served nearly 100,000 women. She also founded the **Association for the Promotion of the Status of Women**, which was a major influence in gaining reform for women's rights in Thailand. In addition, she was one of the founders of Thailand's first college for Buddhist nuns, as well as an AIDS hospice, an orphanage and child care center, a clinic for homeless pregnant women and their infants, and a center for youth.

As a devout Buddhist, Wichiencharoen taught that if we are more concerned about others than we are about ourselves, we will find happiness. She strove never to think ill of others and made special efforts to be friendly toward difficult people. She was especially dedicated to restoring the nuns' ordination to Thailand, hoping to give young women access to Buddhist education and social work, rather than only having the limited choices of domestic oppression, factory work, or prostitution. She was the epitome of Buddhist virtues and was a key player in gaining rights for both Buddhist nuns and laywomen.[61]

Thubten Chodron – Thubten Chodron is an influential American Buddhist nun, a student of the Dalai Lama, and an acclaimed teacher and author. She has also been active in interreligious dialogue with members of the Jewish faith. Chodron has lectured around the world on topics such as peace, human dignity, and the value of serving others. She is a formidable advocate for the liberation of Buddhist women in the West.[62]

Cheng Yen –

>> *Learn more about Cheng Yen and listen to Still Thoughts:*
http://www.youtube.com/watch?v=26rEbWST-hU

Cheng Yen, a Taiwanese nun and one of the most outstanding Buddhist leaders in the area of service today, is the founder of the **Tzu Chi**, also known as **Buddhist Compassion Relief**. Early on, Cheng came to realize that serving in the home was not adequate for her, and that she was called to extend family love to her larger community. Cheng left home and became a Buddhist nun with little means of support. She soon became a popular Buddhist teacher and was eventually hired at a university.

Cheng founded the Tzu Chi organization in 1966, starting in a small temple with a few housewives who saved their grocery money each day, made shoes for babies, and cared for refugees from mainland China. Their commitment was to help the poor and sick in the Buddhist way with love, compassion, joy, and giving. Cheng teaches that when one eye sees, a thousand eyes see; when one hand helps, a thousand hands help.

Soon her organization grew, and today it has volunteers in forty-seven countries in 345 offices worldwide. Tzu Chi is dedicated to serving the poor, especially poor women, and providing disaster relief, environmental protection, health care, and education. Tzu Chi was one of the first organizations to be present after the great Indian Ocean tsunami of 2004 and the devastating earthquake in Haiti in 2010.

Cheng Yen teaches that suffering comes from both material and spiritual poverty. It is her conviction that the lack of loving kindness is at the root of many of today's problems, and she is dedicated to bringing tenderness to as many people as she can.[63]

>> *Google Tenzin Palmo, and view the videos* The Nature of the Mind; Heart Wish; *and* Interview. *See also the book* Cave in the Snow *by Vicki Mackenzie, for an incredible account of Palmo's search for enlightenment while living as a hermit in a cave in the Himalayas.*

SUMMARY

Buddhism began when a young noble left his home and family to seek enlightenment. Eventually he experienced an extraordinary awakening to the truth of how to conquer suffering, live a righteous life, and be freed from endless rebirths and reach nirvana. He was the Buddha, the Enlightened One, and he spent the rest of his life gathering his disciples and teaching his path to enlightenment.

After the Buddha's death, his disciples split into two main groups. The more traditional Theravada Buddhism emphasized the highly ascetical monk's life as the role model and continued to see the Buddha as an exceptional man. The more progressive group, the Mahayana, viewed the Buddha as a celestial being, held the devotional path of the laity to be as important as the monk's path, and held that the Buddha-nature was in all and that all were linked by divine grace. Eventually, as Buddhism spread, other forms developed: Tibetan Buddhism, Pure Land, Nichiren, and Zen.

In modern times, Buddhists were severely persecuted by the Chinese Communists and only just recently have begun to reclaim their traditions in China and India (where Tibetan Buddhists are living in exile). Buddhism has claimed many followers in the West in recent years. Buddhists worldwide are becoming more active in social issues, reinterpreting and

anatta - The no-self

anicca - Impermanence

arahant - Theravada tradition for one who has achieved enlightenment

bhikkhu - Buddhist monk

bhikkhuni - Buddhist nun

bodhicitta - Mahayana term for the Buddha-nature that is in every human being

bodhisattva - Mahayana term for one who delays enlightenment until others have attained it

deva - A deity

dharma - Teaching, truth

dukkha - Suffering

karma - Actions

koan - A Zen term for a puzzle that cannot be dealt with through reason

metta - Loving kindness

nirvana - Liberation from re-birth and suffering

puja - An act of devotion

Sakyamuni - Title given to the Buddha

samatha - Tranquility

samsara - Continuous cycle of rebirth

sangha - The community of monks or nuns

sati - Mindfulness

TEST YOUR LEARNING

1. Is suffering really as prevalent as Buddhists think? Couldn't Buddhist views on suffering lead to pessimism?

2. What is your experience of suffering, and how has such experience affected your outlook? How have you dealt with suffering?

3. How can one work toward a "no-self"? Would that be counter to self-development?

4. Do you see a value of meditation? Do you practice some form of meditation?

5. How does your notion of happiness compare and contrast with the Buddhist notion?

APPLYING BUDDHISM TO WORLD ISSUES (SHORT ESSAYS)

1. Write a brief essay on the value of Buddhist meditation in today's hectic lifestyle.

2. Write a brief essay on how "clinging" to negative thoughts or feelings can cause us personal suffering.

SUGGESTED READINGS

Chappell, David. Buddhist Peacework. Boston: Wisdom, 1999.

Chodron, Pema. Practicing Peace in Times of War. Boston: Shambhala, 2006.

Cozort, Daniel, ed. Journal of Buddhist Ethics. http://blogs.dickinson.edu/buddhistethics.

Dalai Lama. Ethics for a New Millennium. New York: New Riverhead Books, 1999.

Ghosananda, Maha. Step by Step: Meditations on Wisdom and Compassion. California: Parallax Press, 1991.

Hanh, Thich Nhat. The Heart of the Buddha's Teaching. New York: Broadway Books, 1998

———The World We Have: A Buddhist Approach to Peace and Ecology. Berkeley: Parallax Press, 2008.

Harvey, Peter. An Introduction to Buddhism. New York: Cambridge Univ. Press, 1990.

Kaza, Stephanie, and Kenneth Kraft. Dharma Rain: Sources of Buddhist Environmentalism. Boston: Shambhala, 2000.

Puri, Bharati. Engaged Buddhism. New York: Oxford University Press, 2006.

Tucker, Mary Evelyn, and Duncan Ryuken Williams, eds. Buddhism and Ecology. Cambridge: Harvard University Press, 1997.

VIDEOS

Buddhism: Making of a Monk, 1 videodisc. Directed by Richard Keefe. Lawrenceville, NJ : Cambridge Educational, 2004.

Watts, Alan. Buddhism, Man, and Nature, 1 videocassette. Cos Cob, CT: Hartley Productions, 1978.

Hartley, Elda. Buddhism: Path to Enlightenment, 1 videocassette. Cos Cob, CT: Hartley Film Foundation, 1988.

NOTES

[1] *Lord Chalmers, ed.,* Buddha's Teachings: Being the Sutta-Nipata or Discourse Collection *(Cambridge, MA: Harvard University Press, 1997), 193. See Sutta-Nipata, 679-700.*

[2] *Peter Harvey,* An Introduction to Buddhism *(New York: Cambridge University Press, 1990), 16-17.*

[3] *See http://www.accesstoinsight.org/tipitaka/an/index.html, Anguyyara Nikaya, I, 45.*

[4] *Donald W. Mitchell,* Buddhism: Introducing the Buddhist Experience *(New York: Oxford University Press, 2001), 9-10.*

[5] *Thomas Berry,* Buddhism *(Chambersburg, PA: Anima Pub, 1967), 13.*

[6] *I. B. Horner, trans.,* Majjhima Nikaya *(London: Luza, 1955), I,169.*

[7] *Ibid., I, 83.*

[8] *Chalmers,* Buddha's Teachings, *Vol. 37, Book 3, Sutta 9.*

[9] *Harvey, 90ff.*

[10] *Karen Armstrong,* Buddha *(New York: Penguin, 2001), 11-12.*

[11] *See http://www.palikanon.com/namen/maha/mahavagga.htm, Mahavagga, 1:6. 19-22.*

[12] *Ibid., 1:14.5.*

[13] *Chalmers,* Buddha's Teachings, *Book 2, Sutta 4, 267.*

[14] *Gill Farrer-Halls,* The Illustrated Encyclopedia of Buddhist Wisdom *(Wheaton, Illinois: Quest Books, 2000), 14.*

[15] *Chalmers, Book 2, Sutta 1, 225-226.*

[16] *Thich Nhat Hanh,* The Heart of the Buddha's Teaching *(New York: Broadway Books, 1998), 20-22.*

[17] *Chalmers, Book 4, Sutta 2, 773.*

[18] *Ibid., Book 4, Sutta 4,791.*

[19] *Berry, op.cit., 18.*

[20] *Thomas Byrom,* The Dhammapada *(New York: Random House, 1976), 20.*

[21] *Chalmers, Sutta 3, 666.*

[22] *Elizabeth J. Harris,* What Buddhists Believe. *(Oxford: One World, 1998), 53.*

[23] *Maha Ghosananda,* Step By Step. *(Berkley: Parallax Press, 1992), 48.*

[24] *Chalmers, Suttas 146-151.*

[25] *Ibid., Book 4, Sutta 14, 916*

[26] *Shantideva Bodhicaryavatara, 9.56-61,68.73,82*

[27] *See Horner,* Majjhima Nikaya.

[28] *See http://www.accesstoinsight.org/tipitaka/kn/ud/index.html, Udana 8.1-4.*

[29] *T. W. Ryys Davids, trans.,* Milandapancho *(Oxford University Press, 1894), 268.*

[30] *Etienne Lamotte, trans.,* Samdhinirmocana *(Paris: Adrier Maissoneuve, 1935), Book IV, 37.*

[31] *Chalmers, Book 5, Sutta 5, 1074.*

[32] *See http://somewhereindhamma.wordpress.com/.../buddhas-last-words- , Maha Parinibbana Sutta 2, 26.*

[33] *Ibid., Book IV, Sutta 5, 800.*

[34] *Huston Smith,* The Religions of Man *(New York: Harper and Row, 1958), 135.*

[35] *Berry, 64-65.*

[36] *Hanh, 253.*

[37] *Gill Farrer-Halls, 43.*

[38] *Donald W. Mitchell,* Buddhism: Introducing the Buddhist Experience *(New York: Oxford University Press, 2002), 206-209.*

[39] *Harris, 83.*

[40] *Mitchell, 273.*

[41] *Merv Fowler,* Zen Buddhism *(Brighton: Sussex Acad. Press, 2005), 133.*

[42] *See Ven. Gyomay Kubose,* Zen Koans *(New York: Henry Regnery, 1973).*

[43] *Harvey, 139.*

[44] *Harvey, 281.*

45 *Chalmers, Book 1, Sutta 8. 150.*

46 *Mary Evelyn Tucker and Duncan Williams, eds.,* Buddhism and Ecology *(Cambridge, MA: Harvard University Press, 1997), 14.*

47 *Eun-Su Cho, "The Lonely Eco-Fight of Jiyul Sunim," in* Eminent Buddhist Women *(Ho Chi Minh City, 2010),129-132.*

48 *Dhammapada, 41.*

49 *Dhammapada, 405.*

50 *Ranjini Obeyesekere, trans.,* Jewels of the Doctrine, *by Dharmasena Thera (Albany: State University of New York Press, 1991), 98.*

51 *Dalai Lama, 206.*

52 *Ibid.*

53 *Ibid.*

54 *Sutra 707.*

55 *Jamgon Kongtrul Lodro Taye, "The Two Traditions for the Development of the Mind," in* Buddhist Ethics *(Ithaca, NY: Snow Lion Pub., 1998), 172 ff.*

56 *Dhammapada, 201.*

57 *http://www.youtube.com/watch?v=UtkMZMkI5N8*

58 *Ellison Banks Findly, ed.,* Women's Buddhism, Buddhism's Women *(Boston: Wisdom Pub., 2000), 3.*

59 *Diana Y. Paul,* Women in Buddhism *(Berkeley: University of California Press, 1979), 85-86. This book presents a thorough overview of the Buddhist texts on women.*

60 *Hema Goonatilake, "Pioneering Buddhist Women Across Cultures," in* Buddhist Women, *ed. Karma Lekshe Tsomo (Petaling Jaya, Malaysia: Sukhi Hoto Dhamma Pub., 2008), 11-13.*

61 *Karma Lekshe Tsomo, ed.,* Innovative Buddhist Women *(Richmond, Surrey: Curzon Press, 2000), 275.*

62 *See Pema Chodron,* Practicing Peace in Times of War *(Boston: Shambala, 2006), and* When Things Fall Apart *(Boston: Shambala, 2000).*

63 *C. Julia Huang,* Charisma and Compassion: Cheng Yen and the Buddhist Tzu Chi Movement *(Cambridge: Harvard University Press, 2009).*

CHAPTER 4

Judaism

E lie Wiesel, renowned author and Holocaust survivor, was a teenager in the death camp at Auschwitz. He tells of one occasion when
he was taken to a barrack where a group of learned rabbis were
having an intense discussion late at night. They had become disgusted with
the beatings, killings, and burning of innocent people, especially all the
precious children. They had become angry with God for not saving these
innocent people, and they had decided to put God on trial! The trial lasted
several nights with witnesses presenting evidence for the prosecution and
for the defense. At the end there was a unanimous verdict: "The Lord God
is found guilty of crimes against creation and humankind." Then, after a
long silence, one of the rabbis announced: "It is time for evening prayer."
And so the group began the evening service.[1]

>> **See a dramatization of this event:**
God on Trial: http://www.youtube.com/watch?v=I-oNYd23pQk

Wiesel went on to write a play about this incident where he explored some
of the basic questions within his Jewish religion: the meaning of this ancient
covenant that God made with the Jewish people; the devotion and faithful-
ness that Jews have to Torah; the times that they have failed and the many
sufferings they have endured; and God's everlasting commitment to save his
people in the midst of their suffering and failures. In the end, Wiesel admits

that there are no final answers, just more questions raised by a "chosen people," who for 3,000 years have struggled to understand their God and be faithful to his covenant.

What is Judaism? Jacob Neusner, an outstanding Jewish scholar, points out a distinction between Jews as an ethnic group and Jews as a religious people. He observes that not all Jews follow Judaism and that some Jews do not believe in God. Some Jews practice Buddhism or Christianity. To make things even more complicated, Judaism accepts converts, so not all who practice Judaism are Jews!

According to Neusner, Judaism is a religious community that identifies with the biblical "Israel," the group started by the founding fathers and mothers beginning with Abraham and Sarah. Judaism consists of the people who accept the worldview of authoritative holy books compiled in the Torah. And, finally, Judaism refers to the way of life guided by Jewish laws of Sabbath, rituals, the celebration of feasts, and the honoring of life cycles.[2]

In this chapter we are going to study Judaism. We will examine its history—both biblical and post-biblical—its sacred texts and commentaries, its central religious beliefs, and its branches and practices. We will close with a discussion of contemporary Judaism, noting some actions that Jews are taking with regard to ecology, peace, and the women's movement.

THE BIBLICAL HISTORY

Throughout many centuries biblical Judaism was an oral tradition, with the teaching being passed on by parents and leaders at home and in public gatherings and rituals. Around 400 BCE an authoritative set of writings, known as the Torah or first five books, was assembled. A century later the compilation of the *Prophets* was organized. Then, in the late first century CE, the *Sacred Writings* were approved. It was a long process of organizing many narratives and weaving them together into written texts.[3]

Before discussing the biblical history of the Jews, it should be noted that Jews are divided on the historical value of biblical literature. Some traditional Jews accept these writings literally as history, as divinely revealed accounts of God's actions in creating and, then, calling and guiding the Jewish people throughout their history. Others do not count these narratives to be authentic history unless they can be scientifically verified. Some consider the narratives to be historical unless they can be proven to be false.[4] From another perspective, many Jews hold that the question of "what happened?" is not the central question for the Semite mind. Rather, the key question in biblical texts is concerned with faith and religious meaning. From this perspective, the biblical narratives primarily are intended to be teaching stories. Some are completely mythical (the creation stories); others are narratives with some basis in history (the exodus story). But always the intent is to teach the story of God and his unique covenant with the Jewish people.[5]

Creation – The Hebrew Bible opens with narratives about beginnings. In the first of two creation stories, *Genesis 1*, the Creator makes everything in the universe come into being with a mere word. Throughout the narrative, God sees that all of his handiwork is good. God finally creates woman and man in his own image and tells them to increase and multiply. In the second creation story, *Genesis 2*, God makes heaven and earth, forms man from the soil, and breathes life into him. God then places man in a lovely garden and tells him that he can eat of any tree except the tree of knowledge of good and evil. Seeing that man was lonely and discovering that beasts and birds do not suit man, God creates woman from the man's rib.

These myths, in part influenced by Babylonian stories, present a unique perspective on human life. Departing from the pagan notion that humans are part of the divine substance, the Hebrew tradition presents humans as made in God's image, equal among themselves, and accompanied by a God who "walks with them in the cool of the evening."[6] In addition, this is a God who is uniquely concerned about human behavior and expects obedience to his law.

The Origin of Sin – Into this idyllic world of God's creation, sin enters. Eve, the first woman, is persuaded by a snake to eat the forbidden fruit.

She then persuades her mate Adam to do likewise. They are punished with pain and hard work and cast out of the garden. The legacy of sin continues as Cain kills his brother Abel. Yahweh becomes displeased with the sins of the descendants of Adam and punishes them by flooding the earth, saving only Noah and his family. Humankind begins again, as God establishes a covenant with Noah and his offspring, a covenant ever to be symbolized by the rainbow. Humanity has begun, made in the image of God, but now free from control of the pagan gods, free to rebel and be disobedient to God's law.

Abram – One Jewish scholar comments that "Abraham is the first Jew, from whom all Jews trace their descent."[7] Abram, whose name was changed by God to Abraham, is the original patriarch of the Jewish people. (The biblical Jews were referred to as Israelites and after the establishment of the kingdom of Judah were known as Jews.) Genesis tells us that he was a wealthy man who was called by God to leave his land in Ur (possibly in present-day Iraq) and migrate into a foreign land, where God will make him become a great nation and through him bless all communities on earth. Abram had the faith to answer God's call and take his wife Sarah and part of his clan on a distant journey. Abram traveled to Canaan, the land he was given by God. Later, God promised him that his descendants would be as numerous as the stars even though Abram was very old and had no children. In time, his name was changed to Abraham, and God established a **Covenant** with him: "I will be your God and you will be my people." God commanded that Abraham's male children be circumcised. This would be a sign of the Covenant.

Jews still see this Covenant as central to their religion. Its existence promises them salvation from the one God, calls for obedience to his law on their part, and assures them of forgiveness when they fail. This distinctive Covenant later became foundational to both Christianity and Islam. It will be discussed in more detail later when we look at the central beliefs of Judaism.

In their old age, Abraham and Sarah miraculously had a son, Isaac. A defining moment occurs when God asks Abraham to sacrifice this son, only to withdraw the command at the last moment. (This story seems to symbolize how the faith of God's people will be tested, sometimes beyond understanding. It might also be a testament against the pagan practice of human sacrifice.)

Isaac had two sons, Esau and Jacob. Jacob's name would be changed to Israel, and his sons would generate the twelve tribes of Israel. One of Jacob's sons, Joseph, was sold by his brothers into slavery and taken to Egypt where he prospered in the court of the pharaoh. Jacob and his sons, caught in a famine, sought help from Joseph, who took them into Egypt, where they too prospered.[8]

>> **Visit the website Jewishhistory.org** to view Historian Rabbi Berel Wein's *Five Thousand Years of Jewish History parts 1–8.* Watch video clip No. 2 on Abraham and Sarah to learn more about the establishment of monotheism and the moral code of the Israelites.

The Exodus – The *Book of Exodus* tells the story of the Israelites in Egypt. Around 1200 BCE a new pharaoh came into power and condemned the Israelites into slavery. Throughout their period of oppression, the Israelites continued to grow in numbers. Their plight was brought to an end by young Moses, who although a Hebrew, had been raised as an Egyptian in the pharaoh's court. While hiding in the desert after killing an Egyptian who had been abusing a Hebrew slave, Moses heard God speaking from a burning bush and was ordered to return to Egypt and free God's people. Upon his return, Moses had a difficult time persuading Pharaoh to release his people, but after enduring ten plagues and the killing of all firstborn Egyptian children, Pharaoh agreed to let the Hebrews go.

>> **See Passover Story:** http://www.youtube.com/watch?v=B2ePd43aon8

After the Hebrews left Egypt, Pharaoh reneged on his agreement and sent charioteers to recapture them. The story tells of a miraculous parting of the waters of the Red Sea. The Hebrew people crossed the Red Sea safely when the water separated, but the Egyptian charioteers drowned when the waters reunited. This liberation of the Hebrew people is called the **Exodus**. This is the key event in Hebrew history and is a symbolic narrative testifying to God's fidelity to the Covenant with his people. No matter how dark the times, the Jews believe that they will be saved by their God. This great event is still celebrated today by Jews at Passover.[9]

The Israelites were then led to Mount Sinai, where God revealed the Law to Moses and told him to give these rules to the people. This set of laws, known as Torah, includes the Ten Commandments, the rules for remaining faithful to the Covenant with Yahweh, as well as the feasts to be celebrated. The narrative shows how the Israelites had been given physical freedom in the Exodus and now received "spiritual freedom" through the law.

The Israelites continued to wander in the desert for forty years in search of the promised land. When they finally arrived at the land of Canaan, whole generations had died off. Even Moses, the one chosen by God to lead his people out of bondage, was deemed unworthy to enter the promised land and died having only seen the land of Canaan from the top of Mount Nebo. It was left to Joshua to lead the people into Canaan through conquest and then divide the land among the twelve tribes. God continued to direct the people's history.

Even though Moses was not able to enter the promised land, he remains the single most important person in the history of the Jewish people. It was he who led them to freedom from slavery in Egypt and gave them the Commandments at Mount Sinai.

Scholars have mixed views on the historicity of the exodus story. There is some evidence of Semite presence in Egypt and of smaller Semite migrations, but nothing of the proportion in the biblical narrative. Nor is there any archaeological evidence of the destruction of Canaan as described in the story of Jericho. There is a growing view that the Israelites might have been part of those who dwelt in Canaan all along and eventually became dominant through a social revolution.[10]

Israel the Nation – Around the eleventh century BCE, the Israelite tribes were united under a great warrior, Saul, who would eventually become their king. During one battle with the Philistines, a young shepherd named David distinguished himself by slaying the giant Goliath and became a popular warrior. Saul resented David's popularity, and David had to flee. When Saul died in battle, David returned, became king, soon conquered Jerusalem, and established the capital of Israel. David conquered surrounding kingdoms and established the foundation for a dynasty that would last—except for the time of the exile in Babylon—one thousand years. David ruled for forty years (ca. 1013–973 BCE) and is recognized as the greatest king in Jewish history.

David's son, Solomon, would become the next king and would be known for his wisdom and construction of the First Temple in Jerusalem. He is revered as a talented writer and an able leader. During his forty-year reign, he held the northern and southern kingdoms together, established extensive trade routes, and made alliances with many other nations. There is no archaeological evidence of the grandeur of Solomon's kingdom as described in scripture. It would seem that often these Hebrew narratives of earlier events are more theological and symbolic in their value.

After Solomon's death, the two kingdoms, Judah in the south and Israel in the north, went their separate ways, concluding a long period of northern rebellion. Judah would remain the most stable, partially because it was larger and had a population of people with similar cultures. The south was protected by Israel's landmass to the north and by Egypt to the south, with whom they were allied. The north was torn over the worship of Baal in the area and unstable leadership, and was threatened by the ever-increasing power of the Assyrians. In the end, the Assyrians captured the northern kingdom in 720 BCE and transported its leaders and elite to various parts of the Assyrian Empire. The northern kingdom and its people were no more, save for a remnant that would develop in Samaria. They have been described as "the lost tribes of Israel."

During the seventh century BCE, the Assyrian Empire began to crumble and was eventually overtaken by the Babylonians. It was Babylon's King Nebuchadnezzar who invaded Judah, shipped off its soldiers, craftsmen, and leaders to Babylon, and made Judah a puppet state. In 586 he invaded Judah again. This time, he emptied the royal treasury and burned the city and Temple to the ground.

Next, the powerful kingdom of the Persians gained prominence. Their leader Cyrus defeated the Babylonians, allowing many Jews to return to Judah, where they commenced to rebuild their Temple (515 BCE). Others remained in Babylon and the surrounding environs and would later produce their own **Talmud** (commentaries on Torah). Neusner points out that it was during the century after the destruction of the Temple in Jerusalem that basic Judaism as defined by scripture took shape. During that period, the ancient scripture and teachings were gathered to form the Torah or Pentateuch, which later would be joined with the Prophets and the Writings. Thus began the so-called Second Temple Period, which lasted for almost 600 years, until the Temple would be destroyed again in 70 CE by the Romans.

The Second Temple Period – This period in Jewish history, 500 BCE–70 CE, was marked by division and conquest. The Samaritans, converts who had been practicing what they considered to be traditional Judaism, became alienated from the Jews from the southern kingdom who were returning from exile. The Samaritans did not recognize the Temple in Jerusalem nor the priesthood that served there. (These would be the Samaritans that Jesus encountered in the gospel stories.) Judaism would become reestablished by those returning from exile. The Temple in Jerusalem would be the central place of their sacrifice and worship, the priesthood would be held in high esteem, the wisdom of prophets would be revered, and the instruction of the sages would be carefully observed.[11]

There would be more devastating conquests of the Jews. The most significant would be by Alexander and the Greek Empire (330–323 BCE), which would exert heavy Hellenic influence on Jewish thought and way of life. The Hebrew scriptures would be translated into Greek (the Septuagint) in the early second century BCE, and there were attempts to introduce Greek rites into the Temple, which as we shall now see, would be, in part, grounds for rebellion in the second century BCE.

In 167 BCE there was a rebellion in Judah against the Seleucid Empire that was dominating Judah and shaping the culture with Greek customs. Judah became split into those that supported the Hellenistic influence and those who resisted it. At one point, edicts were published forbidding the practice of the Jewish religion, and a revolt was led by families known as the Hasmoneans. After much struggle, the Maccabee family prevailed and established a strong Jewish dynasty that would last until the Romans came to dominate Judah.

POST-BIBLICAL HISTORY

The next conquerors of Israel were the Romans (63 BCE), who had been originally invited to stabilize the area. Once there, they decided to stay in order to gain control of the land and resources, as well as gain strategic control of the eastern side of the Mediterranean. The Romans reorganized Palestine, gave the prime property to Phoenicians and Syrians, and reduced many Jews to tenant farmers in the hinterlands. Around 37 BCE Rome appointed Herod as a Jewish king and collaborator in charge of the area, now considered to be a Roman client state. Herod built whole cities and monuments to himself and to the Romans and reconstructed the Temple into a magnificent complex, where millions of Jews living in the Diaspora (scattered colonies of Jews outside Palestine after the Babylonian exile) throughout the Middle East and even as far as Rome would come on pilgrimage to worship at the Temple.

Ancient Jewish Sects – By 150 BCE many Jewish religious groups had developed in Israel and most of these appear in the Christian gospels. The **Pharisees** were members of a prominent sect. For Jews in the Diaspora who did not have access to the Temple, the Pharisees offered leadership in the local synagogues. They strove to separate themselves from impurity, to provide the common folk with teaching, and to set good examples for living as virtuous Jews. Their mission seems to have been to improve religious observance in Judah by teaching and interpreting Torah. The Pharisees accepted both the oral and written interpretation of the Law. They believed in the resurrection of the dead and that the world was guided by God's providence. The two key Pharisees in the first century BCE were Hillel and Shammai. Both had a great deal of influence on teachers following them. Josephus, a famous Jewish historian, offers these eyewitness comments on the Pharisees: "The Pharisees simplify their standard of living, making no concession to luxury....They show respect and deference to their elders.... They hold that to act rightly or otherwise rests indeed with humans, but that in each action fate cooperates. Every soul, they maintain, is imperishable, but the soul of the wicked will suffer eternal punishment."[12]

The **Sadducees** were members of the wealthy aristocratic class that included the chief priests, elders, and lay nobles. They controlled the **Sanhedrin**, the councils found in every city in the land of Judah, as well as the **Great Sanhedrin**, the Supreme Court in Jerusalem. Most of the Sadducees supported the Roman occupation because they received lucrative appointments from the Roman authorities. In contrast to the more open Pharisees, the Sadducees were conservative and did not believe in innovations like the oral tradition, God's providence, or the resurrection of the dead. Josephus has this to say about the Sadducees: "They own no observance of any sort apart from the laws; in fact, they reckon it a virtue to dispute with the teachers of the path of wisdom....The Sadducees, the second of the orders, do away with fate [providence] altogether, and remove God beyond, not merely the commission, but the very sight of evil....As for the persistence of the soul after death, penalties in the underworld and reward, they will have none of them."[13]

The **scribes** ranged from being copyists of Torah to learned scholars, whose power came from their influence in interpreting the tradition and in forming future scholars. The scribes felt privileged because not only did they know the tradition, but they could also contribute to the development of the oral tradition.[14] Pharisees recognized their authority to create oral tradition, while Sadducees held only to the written law.

The **Essenes** were members of a reclusive Jewish sect. They lived in communities in the desert and towns, cut off from traditional Judaism. They did not recognize the Temple or its priesthood, and they led strict, ascetic lives, expecting that the Messiah would come to them in the end time. In 1948, Essene documents, which came to be part of a collection called the Dead Sea Scrolls, were found in a cave on the northwest shore of the Dead Sea. Since that time, the study of the scrolls has offered extremely valuable insights into Judaism during the first century BCE. These documents, along with translations, were digitized and made available for viewing online in 2010.[15]

Another group, the **Zealots**, did not accept the Roman domination of Judah. The Zealots were dedicated to terrorist acts against the Romans and anyone who cooperated with them. In the latter part of the first century CE, they became a formal group and led the rebellion against the Romans. The Zealots, Essenes, and Sadducees came to an end as factions when Jerusalem and the Temple were destroyed in 70 CE.

Finally, in the first century CE there appeared the Jesus movement, a small band of disciples who believed that Jesus—a Jew, former carpenter, wandering preacher, and miracle worker—was the Son of God, the Messiah. They had experienced his death and resurrection from the dead and preached his gospel to both Jews and non-Jews. One of these disciples, a former Pharisee named Paul, was dedicated to taking this message to **Gentiles**, non-Israelite peoples or nations, and the movement spread across Asia Minor into Greece and even Rome.

>> **Watch Judaism Part 1** http://www.youtube.com/watch?v=NlkdEr3j5kU&feature =PlayList&p=1F352FDD12D732FE&playnext_from=PL&index=19 **and consider the complexity of the Jewish faith and the unique aspects of Judaism.**

The Rabbinic Period – During the first century CE, there were uprisings among the Jews in Judea and Egypt against the Romans, and ongoing feuds between the pro- and anti-Roman Jews, as well as between the Greeks and Jews. For a time, the Romans did not have the personnel to control the area, and the Zealots were able to take over Jerusalem. In 69 the Emperor Vespasian put his son Titus in charge of the Judean campaign. In 70 Titus came with a huge force and destroyed Jerusalem with its Temple and slaughtered the residents.

A sizable number of Jews moved north to Yavneh, and under the leadership of Yohanan ben Zakkai a number of scholars began the long process of determining the canon of the scriptures and organizing rituals, daily prayers, and the Passover Seder meal. Since there were no longer Sadducees nor priests, the Pharisees now led the Sanhedrin, and rites were formulated to ordain rabbis, who would now function as the Jewish teachers.

Jewish rebellions continued in parts of the Diaspora for the next decades. In 132 Simon bar Kokhba led a massive rebellion in Judea against the Romans. The Jewish rebels were defeated in 135. For a time, Jews were not allowed in Jerusalem. It did not take long, however, for the Jews to reconcile with the Romans, especially under the leadership of Rabbi Judah the Prince, who was well-regarded by the Roman leaders. The Jewish center was now moved to Galilee, where great academies were established and work was begun on the Mishnah, which was the compilation of the oral teachings that had been passed on through the ages.

Rabbinic Judaism – Even though Jerusalem and the Temple had been destroyed, the Jews were prepared to develop new forms of Judaism. Already half the Jewish population lived outside of Judah—in Egypt, Syria, Greece, North Africa, Spain, Gaul, throughout the Middle East, and even in Rome. Judaism had for centuries adapted to other cultures and had established synagogues for prayer and instruction. Judaism was no longer dependent on a national homeland and could exist independently anywhere.

The Pharisees and scribes led the way in shaping rabbinic Judaism. Neusner points out that they "carried forward the oral traditions that were associated with the Torah and formed the community of Judaism that observed the written Torah in accord with the oral traditions....Its possession of written and oral traditions earned for it the title, 'the Judaism of the dual Torah,' written and oral."[16]

The main commandment for rabbinic Judaism would be the study of Torah. The Torah shows the way to live a righteous life before God. Holiness that in the past was associated with the Temple was now centered in the life of the people. The sanctification of Israel now transcended buildings and sacrifices, for it was now the people who were to be the "medium and the instrument of God's sanctification."[17] Now, no longer needing the intervention of priests, both women and men could learn the mind of God, both individually and in community. The **rabbi**, a position never mentioned in the Hebrew Bible, would become the new teacher and leader of the people. The ancient teacher Simeon the Righteous had taught that the world stood on three pillars: "Torah, worship, and loving kindness." Now there would be but two pillars, Torah and loving kindness, but Judaism would still stand strong.

With the Temple gone, atonement for sin could no longer be gained by animal sacrifice through priests. The notion of **teshuvah**, sincere repentance, which had been given importance in Leviticus, a book in the Torah, was now placed at the center of Jewish worship. Teshuvah means "return" or "respond" and here refers to returning or responding to God after transgressing. With no altar to use for atonement, the sinner must return to God in his or her heart, and once again find at-one-ment or reconciliation with the Lord. Teshuvah would become the central theme of the high holy days between Rosh Hashanah and Yom Kippur (which will be discussed further in the section on feasts).

Torah in rabbinic Judaism was to be loved and embraced. Its laws were there to transform the Jew into the embodiment of what God wills—a living law. Torah was viewed as the creative Word, and it was thought that without it the world would plunge back into the chaos that existed before creation. The rabbinic period produced additional "learning" about Torah in the **Mishnah** (ca. 220 CE). The oral discussions and rulings of scholars and rabbis were gathered and put into volumes that would assist the Jewish communities in living their everyday lives in line with Torah.

With the Temple gone, importance was now given to the synagogue, which earlier in the Diaspora had been local meetinghouses where readings and some prayers were conducted. Now, Jews would gather in the synagogues for "worship of the heart" and prayer, both individual and communal. With the rabbi as leader, the community would celebrate God's daily renewal of creation. The central prayer would become the Amidah (the standing prayer), which praises God and petitions God and gives thanksgiving. Throughout, there is recognition of God's blessings. The prayer would end with a request for peace. Rabbinic Judaism also stressed deeds of loving kindness, based in part on the command in Leviticus to "love your neighbor as yourself." The rabbis stressed that such love is at the very heart of Torah and is the way to atonement.

For rabbinic Judaism, the Sabbath now became the celebration of sacred time and space, and its observance was deemed crucial to salvation. Shabbat was to be a time free from all work, a time to delight in God's presence, re-create the self, and to taste the joy of anticipating the coming world. As one rabbi, Ahad Ha'am (1856–1927), said: "More than the Jews have kept the Shabbat, Shabbat has kept the Jews."[18]

Facing Christianity – Rabbinic Judaism provided a firm foundation for dealing with the challenges the Jewish people would face in the coming centuries. The first of these challenges was the appearance of Christianity in the first century. Originally a Jewish movement, Christianity later became predominantly a Gentile movement, and hostility developed between the two faith groups.[19] The Jews often viewed Christians as idolaters for worshiping Jesus as God, and the Christians frequently portrayed Jews as those who rejected and killed the true Messiah. For Christians, Jesus Christ and the gospel teachings were the way to holiness and resurrection, while for the Jews, Torah was the way. Even though Paul moved Christian conversion toward the Gentiles (non-Jews), a number of Jews still converted.[20]

In 380 CE Christianity became the official religion of the Roman Empire. Judaism lost the protection and tolerance it had received under Roman rule, and became marginalized. Though Jews were banned from government positions and were second-class citizens, they were tolerated because of their connection to Christianity and with the hope that they would see the light and convert.

But Rabbinic Judaism enabled the Jews to remain strongly bonded as a people, and centers of practice and scholarship flourished. In fourth-century Galilee the teachings of the rabbis were gathered to further develop the Mishnah, which would ultimately become the Palestinian Talmud (commentary on scripture). In Babylon, the Jewish community produced accomplished scholars. Though the Jews were subjected to Persian law and later persecuted by Christian leaders, in the fifth century these scholars were able to complete extensive commentaries on Mishnah and produce the Babylonian Talmud, which was four times longer than the Talmud of Palestine.[21]

Christian leaders usually required the Jews to live in contained, enclosed communities. In Italy, where Jewish communities once flourished, Jews were driven off farmlands, their synagogues at times burned. Even though the papacy forbade forced conversions, Jews were expected to and often pressured to convert. In spite of the pressures and persecution, the communities were able to thrive and produce excellent scholars and scientists. In the Byzantine Empire, Jews were treated much more harshly and frequently forced to convert. In Spain, the Jews were often persecuted by Christians and were forbidden to eat or live with Christians. Under the later Visigoth kings, who had converted to Christianity and conquered Spain, Jews were forced to convert, and circumcision and Sabbath observance were forbidden.

Judaism Meets Islam – Judaism often fared much better under Islamic rule than it did under Christian rule. Muhammad was at first friendly toward the Jews, thinking that his revelations from Allah served as a corrective to the Jewish tradition and the Jews would convert. When they refused, Muhammad became alienated from Judaism. After the death of the prophet, Islam made many conquests and established an empire that extended from the eastern Byzantine Empire and included Syria, Israel, Egypt, Iraq, Persia, Spain and parts of continental Europe. Much of the Jewish community now fell under Islamic rule.

Under Islamic rule, the Jews were still second-class citizens. Under the Pact of Umar (800 CE), they were not allowed to build new houses of worship, make converts, or carry weapons. Jews also had to wear distinctive clothing to mark their identity, pay special poll and land taxes, and accept the authority of the Islamic state. At the same time, Jews were under no pressure to convert as they were in Christian lands. Jews were also guaranteed religious toleration, juridical autonomy, and exemption from the military. Under these conditions, Judaism was often able to flourish in Islamic lands. Jews prospered in the urban areas in many crafts, became key figures in trade, and were employed as financiers for the nobility.[22]

THE MEDIEVAL PERIOD

By the medieval period Jews had spread to many parts of Europe. In Islamic Spain, Jews experienced a Golden Age of literature, poetry, philosophy, and religious writing. In the eleventh century, Christians began to conquer the Muslims in Spain. As they took parts of the kingdom, they dominated the Jews. At first, the Jews were tolerated because of the high positions they held and the high taxes they paid. By the thirteenth century anti-Jewish attitudes appeared, and by 1492 the Jews were expelled from Spain. As they dispersed, they became known as the Sephardic Jews, and they moved on to North Africa, the Balkans, and the Middle East.

Jews in Italy were given some protection in the seventh century by Pope Gregory I. In the ninth and tenth centuries, southern Italy was controlled by the Byzantine rulers, who had little use for the Jews and tried to force them to convert. Still, Talmudic academies and Jewish scholarship flourished in the early Middle Ages. Jews gained prominence in the fields of medicine, astronomy, and classical languages. They were protected by the medieval popes until the Lateran Council in 1215, when Jews were lowered to the status of servants and forced to wear badges identifying themselves as Jews. In the thirteenth century the Inquisition, which will be discussed in more detail later, brought about the torture and execution of many Jews.

Jews came to England with the Normans in the eleventh century. They often prospered as moneylenders (Christian law forbade usury), merchants, and bankers. In the twelfth and thirteenth centuries the tide turned against them: funds were confiscated and their property was taken away. In 1190 many Jews were massacred in York by Crusaders and other nobles, all in debt to the Jews and finding this an easy way to "settle up."

Jews came to Germany in the tenth century and, again, established themselves as successful moneylenders and merchants. In the fourteenth century prejudice began to build, and there were persecutions and massacres in Würzberg, Nuremberg, and in Bavarian towns. In 1348 a plague that became known as the Black Death struck the area, killing nearly half the population. Some maintained that the Jews were responsible for causing the plague by poisoning wells. In the thirteenth and fourteenth centuries there were many expulsions of Jews from Germany and Western Europe. They migrated to the Slavic countries, especially Poland. These European Jews are referred to as Ashkenazi Jews. At present, most North American Jews and nearly half the Jews in Israel are descendants of the Ashkenazi Jews.[23]

Outstanding Medieval Scholars – In spite of the persecutions that the Jews were subjected to during the medieval period, they were still able to establish strong academies in their communities and produce brilliant scholars and writers. Two of the most renowned were Judah Halevi (ca. 1086–1141) and Moses Maimonides (1135–1204).

• Judah Halevi

Judah Halevi was an accomplished poet and songwriter. Some of his songs are such classics that after many centuries they are still sung by Jews today at feasts. In his *Poems of Zion*, Halevi shows his displeasure with the Diaspora and his longing for the homeland of Israel. *Kuzari* is his wonderful prose work where he celebrates the greatness of Judaism and his belief that Jews are God's instruments for enlightening humanity. His words give you a sense of his longing to return to the Holy Land of his ancestors: "If only I could roam through those places where God was revealed to your prophets and heralds. Who will give me wings, so that I may wander far away? I would carry the pieces of my broken heart over the rugged mountains. I would bow down, my face on your ground; I would love your stones; your dust would move me to pity. I would weep, as I stood by the ancestors' graves....The air of your land is the very life of the soul....It would delight my heart to walk naked and barefoot among the desolate ruins where your shrines once stood."[24]

• Moses Maimonides

Most notable during this period is Moses Maimonides (1135–1204), one of the world's greatest philosophers. Maimonides spent his childhood in Córdoba, Spain, but had to flee from Muslim extremists when he was a teenager. The family eventually settled in Egypt, where Maimonides served as a physician in the Muslim court. His greatest work was his fourteen-volume Mishnah Torah, which is an extensive summary of Jewish law and commentary on a vast number of topics. His masterpiece entitled T*he Guide for the Perplexed* provided a monumental philosophical interpretation of the biblical presentations of God, creation, divine providence, the prophets, and many other topics. For instance, in his discussion of charity, he described in great detail the eight dimensions of charity. In philosophy, Maimonides did for Judaism what Thomas Aquinas did for Christianity: analyze the teachings from the point of view of Aristotelian philosophy. Maimonides was convinced that reason and religion were compatible and that reason could be used at the service of religion. At first Maimonides' views were threatening to many Jewish leaders, but eventually his works were incorporated into the Jewish canon of literature.[25] He has been a significant influence in persuading Jews that one can be religious and still be seriously devoted to developing one's intellectual powers. From his vast study of the Jewish tradition, this is his clear and precise summary of the Jewish Articles of Faith:

- belief in the existence of God
- belief in God's unity
- belief in God's in-corporeality
- belief in God's eternity
- belief that God alone is to be worshiped
- belief in prophecy
- belief in Moses as the greatest of the prophets
- belief that the Torah was given by God to Moses
- belief that the Torah is immutable
- belief that God knows the thoughts and deeds of men
- belief that God rewards and punishes
- belief in the coming of the Messiah
- belief in the resurrection of the dead[26]

The Mystical Tradition – Mystical experience seems to have always been part of the Jewish tradition. The mystical movement points to the experience of God in the burning bush (Exodus 3:1–21), and the marvelous visions of Ezekiel of flashing fire, winged creatures, wheels within wheels spinning in all directions, and the sapphire throne above the firmament (Ezekiel 1:1–28). There are accounts from the early rabbinic period of rabbis penetrating the heavenly mysteries as they studied Torah.[27]

It was in the thirteenth and fourteenth centuries that the mystical tradition came to the fore and was associated with the term "kabbalah," the Hebrew word for "that which is received." Kabbalah was a movement in southern Spain that developed techniques for meditation, including breathing exercises and visual imaging. The goal was to elevate the self into a state of spiritual ecstasy and be in union with God. Often the Hebrew letters were used, taking the position that these were from the sacred words used by God in bringing about the Creation. God's holy names were also used during these meditations.[28] In 1290 a text called *Zohar-Brightness* became the primary source for kabbalah. Purportedly it was an ancient rabbinic text, but in fact, it seems to have been written around 1286 by Moses de Leon, a Spanish Jew. When the Jews were expelled from Spain in 1492, the movement spread into Europe and the Middle East.[29] In the sixteenth century new strands of mystical teaching were woven into the kabbalah, and the mystics taught that one who lives a holy life can actually repair the earth and bring about the coming of the Messiah. This messianism left the kabbalists open to accept Shabbetai Tzvi who declared he was the messiah. When Tzvi became a Muslim and was discredited, the kabbalists lost support. The movement rebounded in the eighteenth century when Hasidism integrated kabbalah mystical teachings into the Hasidic emphasis on joyful living and religious experience connected with music, dancing, and storytelling. Tragically, this impetus for kabbalah was lost when the Nazis and Russian Communists all but annihilated the Hasidic communities in the mid-twentieth century.

Today there is a revival of kabbalah among Jews as well as non-Jews. Some Jews in Israel turn to mystical prayer as a relief from politics, economics, and religious legalism. The contemporary interest in spirituality, meditation, and contemplation has moved some in Europe and the United States toward kabbalah.

The Inquisition – In the thirteenth century the infamous Inquisition attempted to rid the Roman Catholic Church of heretics. Suspects were handed over to civil authorities and tortured. If they repented of their heresy, they were imprisoned; otherwise, they burned at the stake. By 1250, Talmuds were being burned in France, and then Spain became the center for the burning of Jews themselves. Special targets were the so-called Marranos (the Spanish word for "pig"), Jews who converted to Christianity, often under duress, and yet continued to practice Judaism secretly. Once discovered, the Marranos were tortured and, if unrepentant, were burned at the stake. Many fled to Muslim countries or to places like Holland or Greece where they could find more acceptance and religious freedom.

THE POST-MEDIEVAL PERIOD

In the centuries following the medieval period most Jews lived in Eastern Europe or the Ottoman Empire (Muslim areas of Turkey, Greece, Egypt, and the Middle East). Large numbers had fled there from the Inquisitions in Spain and Portugal and the expulsions from England and other countries in Europe. Once resettled, they began to flourish, working for the nobility in financial matters, developing academies, and renewing interest in mysticism and the coming of the Messiah. In this area, there was a major and widespread stir among Jews when in 1665 Shabbetai Tzvi, a Jew from Palestine, proclaimed that he was the Son of God and the Messiah. Many Jews quit their jobs and began fasting and observing extreme ascetical practices, preparing for the end time. Tzvi was arrested by the Muslims in Turkey and told to convert to Islam or be executed. Much to the dismay of his followers, Tzvi became a Muslim. Most of his followers drifted away, but the movement hung on for some time before it died out.

During the sixteenth and seventeenth centuries in Europe the Jewish fate often moved from success to persecution. In Poland, for instance, the Jews did well as scholars or estate managers for the nobility and merchants. In 1648, however, the Ukrainian Cossacks turned on the Polish nobility along with the Jews that served them. Numerous Jews were massacred. In Germany the Jews survived many persecutions but were able to establish themselves in banking and trade. With the advent of the Protestant revolt, the tide began to turn against them. Martin Luther (1483–1546), a prominent leader of the movement, was at first well-disposed toward the Jews, hoping for their conversion. When that did not happen, Luther turned on the Jews with a viciousness that remained long influential in Germany. He wrote: "No one wants them. The countryside and the roads are open to them; they may return to their country when they wish; we shall gladly give them presents to get rid of them, for they are a heavy burden on us, a scourge, a pestilence and misfortune in our country….First their synagogues should be set on fire, and whatever does not burn up should be covered or spread over with dirt, so that no one may ever see a cinder or a stone of it. And this ought to be done for the honor of God and Christianity…."[30]

The Jewish Enlightenment – The European age of enlightenment or age of reason, a time of great advancement in scholarship and science, had a marked effect on the way people viewed religion. Religion could now be put under the microscope of reason, and many of its "revealed" truths would be challenged. All this, of course, affected many Jewish scholars. In Holland, where the Jews has been given much freedom for education and research, Uriel Acosta (1590–1640) took the position that the Torah was of human origin and not divine revelation. Baruch Spinoza (1632–1677), who would go on to become a renowned philosopher, maintained that much of the Torah was not composed by Moses but by later generations as an ethical document rather than religious revelation. (Spinoza was formally excommunicated from the Jewish community.)

The greatest of the Jewish thinkers of this time was Moses Mendelssohn (1729–1786). He pursued secular studies in Berlin. In his work, he argued that reason on its own could discover the existence of God, the process of creation, divine providence, and the immortality of the soul.

Mendelssohn was a strong advocate for freedom of religion for all people and a strong critic of coercion with regard to religious dogmas. He maintained that Torah held laws and commandments, not dogmas like the Christians were forced to believe. He said: "I believe that Judaism knows of no revealed religion in the sense in which Christians understand the term. The Israelites possess a divine legislation—laws, commandments, ordinances, rules of life, instruction in the will of God as to how they should conduct themselves in order to attain temporal and eternal happiness…but no doctrinal opinions, saving truths, no universal propositions of reason."[31]

Mendelssohn encouraged his Jewish people to move out of the ghettos and to become sophisticated and learned members of the society, all the while remaining devoted Jews. He wrote: "Adopt the customs and constitution of the country in which you live, but also be careful to follow the religion of your fathers."[32]

By the late eighteenth century there were 2 million Jews in Christian Europe. In England and Holland they were relatively free from restrictions. In Central Europe they were often seriously restricted and had to wear special signs and hats to identify themselves as Jews. The American and French revolutions gave impetus to more freedom for Jews in Europe. New laws were passed that allowed Jews more access to jobs and education; ID badges and hats became a thing of the past. Many Jews were even allowed to become citizens where they lived.

Germany was an exception because the government did not resonate with the new spirit of emancipation. Jews were still viewed as "Asian aliens," deprived of civil rights, and pressured to convert to Christianity.[33] Then, in the nineteenth century, Germany experienced a new spirit of liberalism. In 1869 the Jews in Germany and Austria were emancipated and given the rights of citizenship.

By the end of the eighteenth century in Eastern Europe, partitions of Poland left most of the Jews in that area subject to the authority of the Russian Czars. Jews were restricted as to where they could live, pressured to convert, and often subject to military service and deportation. Under Alexander II (1855–1881) the Russian government became more liberal. Jews were allowed to move to large cities such as Moscow and St. Petersburg and enter the schools and professions. After Alexander II was assassinated in 1881, the Jews were once again persecuted and forced to live in restricted areas. In 1891 many Jews were expelled from Moscow, and in the early nineteenth century there were more persecutions of the Russian Jews.

In the eighteenth century an enthusiastic Jewish movement arose in Poland and Lithuania known as Hasidism. The disciples of this movement were characterized by intense emotion for Torah, devotion to their leaders, and a spirit of joy and celebration. As mentioned earlier, the Hasidim were strongly attracted to the Jewish mystical tradition and kabbalah due to its emphasis on ecstatic experience of the divine. By the 1830s large numbers of Jews in Southern Poland, Ukraine, Hungary, and Lithuania were Hasidim. Many of those came over to the United States in the late nineteenth-century immigrations.

THE MODERN PERIOD

In the nineteenth century the growth of nationalism and liberalism presented Judaism with new freedoms as well as challenges. Rather than existing as segregated groups within countries, Jews were given the opportunity to become citizens of individual nations such as France, Germany, Poland, and the United States, as well as have more opportunities to gain education at universities. While many Jews rejoiced at these gains, they disagreed profoundly on how they could adapt to the differing national cultures and still preserve their identity as practicing Jews. There were calls to modernize the worship services, change the dietary practices, and even look at the Bible in light of modern scholarship. The differences would cause unprecedented divisions in Judaism and separate the religion into various divisions: Orthodox, Conservative, Reform, Reconstructionists, and others.

Orthodox Judaism – Orthodox Judaism maintains that Torah was divinely given to Moses and is not subject to change. The laws of Torah, both written and oral, are from heaven, and the faithful Jew must follow them. Therefore, Jewish law should prevail in the life of Jews, not the laws of nations or secular society. Orthodoxy is conservative and, therefore, dedicated to conserving the traditions of Hebrew language, law, lifestyle, dietary laws, ceremonies, and Sabbath.[34] Modern secular culture is usually viewed to be a threat to all these traditional observances and to the Jewish identity. At the same time, Orthodox Jews attempt to adapt to the country in which they live and to the local customs. Since the rise of Jewish Orthodoxy in the nineteenth century, there has been a diversity of views on synagogue ritual, the value of mysticism, Zionism, leadership roles, and cultural adaptation.

Reform Judaism –

>> *Learn more about the Movement for Reform Judaism:*
http://www.youtube.com/watch?v=GwdB3UL5JTMhttp://www.youtube.com/
watch?v=GwdB3UL5JTM **and consider ways they are reaching out to Jewish**
youth to engage them in faith practice.

Reform Judaism highlights change and adaptation as its hallmarks. Members of this movement no longer hold that Torah was divinely given to Moses and accept the human authorship of Torah, as well as the literary and mythical quality of scripture. Torah is still held in high esteem, but it is understood as having been influenced by past cultures and therefore adaptable to changing times. Reform Judaism began in nineteenth-century Germany, and the followers who worshiped in German rather than Hebrew preferred to call their places of worship temples, rather than synagogues. Reform Judaism acknowledges the ethical values and ideals of Torah, but does not see the need for a literal interpretation, nor for following specifics of the past with regard to clothing, diet, or everyday living.

There is a diversity of views within the Reform movement. Some have moved toward a more traditional perspective in order to maintain identity, especially in the wake of the horrors of the Holocaust and the rise of the nation of Israel. There is a renewed interest in the Hebrew language and in studying traditional views. Many in the Reform movement emphasize the prophets and are drawn toward social justice. Once excluded in Israel, Reform Judaism now has a recognized presence there.

Conservative Judaism – Conservative Judaism tries to walk the middle way between the Orthodox and Reform. Largely an American movement, it attempts to hold on to the tradition, adapt to culture and modern biblical scholarship, and at the same time avoid changes that they consider to be excessive in Reform Judaism. They accept the fact that Judaism has and still does evolve historically, but at the same time they hold firm to Jewish law, carefully adapting to the changing times.

Reconstructionist Judaism grew out of the Conservative movement. It does not see God as a supernatural being, but as the animating force in the universe. The movement views Judaism as a civilization and interprets Jewish beliefs in secular terms.

EVENTS SHAPING MODERN JUDAISM

Two major events have shaped modern Judaism: the **Holocaust** (Shoah), where six million Jews, including 1.5 million children, were killed by the Nazis during World War II, and the establishment of Israel as a Jewish State in 1948.

>> *View the Holocaust album at* http://www1.yadvashem.org/yv/en/holocaust/index.asp *and reflect on all that was "exterminated" in these camps.*
Read a Holocaust survivor story
http://www.ushmm.org/wlc/en/article.php?ModuleId=10007523

The Holocaust – The oppression of Jews in twentieth-century Germany began in 1933 when Adolf Hitler's Nazi party blamed the Jewish people for the devastation of Germany during and after the First World War. The Nazis intensified an anti-Semitism that had been developing in Russia, Germany, and Eastern Europe since the nineteenth century. The Nazis decided that Jewish property and possessions were to be confiscated, their businesses boycotted, and their positions within government and universities eliminated. Their rights as German citizens were revoked, and they became subjects of the state. The Jews were to become the scapegoats for all the ills in Germany following World War I. Shortly after the Nazis invaded Poland in 1939 and began the Second World War, there were efforts to deport the Jews, then to crowd them into ghettos. It soon became clear that there were too many Jews in the ghettos and sustaining them would mean depriving the German people of resources. The many efforts to shoot thousands of Jews and push them into burial ditches had become difficult and often traumatic for the soldiers doing the executions. A conference of Nazi officers and officials was held in Wannsee, near Berlin on January 20, 1942. The agenda was to find "The Final Solution to the Jewish Question." The minutes reveal that it was decided that camps were to be constructed where the Jews would be systematically gassed and their bodies incinerated. Hundreds of thousands had already been killed by death squads. Now the Reich would gear up for a more "efficient" way to industrially eliminate all the Jews of Europe.

Visitors to Auschwitz-Birkenau, the largest camp, located in Poland, are struck by its vast expanse. Auschwitz-Birkenau was actually three camps: a prison camp, an extermination camp, and a slave-labor camp. All that remains of the hundreds of barracks, where people were housed like animals, are the thin, brick chimneys. The Birkenau gas chambers and cremation ovens were blown up by the Nazis before they retreated and are now only a pile of rubble. Located at a railway junction with forty-four parallel tracks, trains consisting of freight and cattle cars brought Jews from all over Europe to the camp. The men who had some strength left were put to work until they were too weak to be useful, and then they were killed. The women and children, along with the elderly, sick, and handicapped, were immediately taken to the gas chambers, having been told that these were bathhouses. Once locked in, gas pellets were released through roof vents, and the fatal gas filtered to the people below. In a few agonizing minutes, it was all over, and the bodies were dragged to corpse cellars until they could be put into the cremating ovens. Four thousand people could be disposed of in one single day. Visitors to Auschwitz-Birkenau silently walk past rooms filled with the victims' suitcases, thousands of eyeglasses, and huge piles of shoes. The tiny shoes of the children are the hardest to observe.

By the end of the Holocaust more than 12 million people died in concentration camps. In addition to their plans to annihilate the Jewish race, the Nazis victimized political dissidents, Gypsies, non-Jewish Poles, the disabled, the mentally ill, the emotionally disturbed, Jehovah's Witnesses, homosexuals, and prisoners of war.

Not much was said about the horrors of the Holocaust immediately after the Second World War ended. For the survivors, it must have been difficult to describe. In 1960 Adolf Eichmann (1906–1962), the man who supervised the deportations of Jews to the death camps and one of the masterminds of the Final Solution, was arrested while hiding in Argentina, put on trial in Jerusalem, and executed. That event, added to Jewish concern for Israel during the Six-Day War with the Arabs in June of 1967, seemed to stimulate memories of the Holocaust and provoke discussion. Many survivors gave testimonies and Holocaust museums were opened around the world. The two most famous are Yad Veshem in Jerusalem and the United States Holocaust Memorial Museum in Washington, D.C.[35]

Many Jews have struggled to comprehend the horrors of the Holocaust. Some reflected that perhaps this was a punishment, and yet had to ask how anyone could have deserved such cruelty. Others thought that God was dead and that the Covenant was over. Some asked why so many people, and perhaps even God, were silent and turned their faces from this horror. Indeed, many Jews kept the faith even in the face of the horrors they experienced, possibly believing that God was with the victims as they walked to the gas chambers, and that God helped the survivors who were blessed to live on and restart their lives.[36] Many Jews stood in faith before God like Job, not able to understand the mystery of why bad things happen to good people, and yet willing to live faithfully with these mysteries.[37]

Zionism and Israel – After two thousand years, the nation of Israel is again located in the land of their ancestors and once again controls Jerusalem. Interest establishing a Jewish homeland began in the nineteenth century with Theodor Herzl (1860–1904), the founder of the secular, political **Zionist movement**, a response to the anti-Semitism prevalent in the period. The Zionists sought to establish an independent Jewish state in the ancient homeland of the Israelites. After World War I, Palestine was under British control, and during World War II Jewish immigration began to increase in Palestine. The British attempted to restrict this immigration, but the near extermination of the Jews during the Holocaust turned public support to the Jews. In 1947 the United Nations' partition plan divided Palestine into separate Arab and Jewish states, with Jerusalem under U.N. control. In 1948 the British left Palestine, and the State of Israel was proclaimed. The surrounding Arab states rejected the U.N. partition and invaded Israel. An armistice was declared in 1949, and Israel vastly increased their territory as a result of the war.

The Holocaust and the establishment of the State of Israel have given many Jews a renewed identity. Once again they had experienced exile and redemption. Once again many Jews believed that the Covenant with their loving God had been affirmed. As Jacob Neusner said: "A Judaic system, the Judaism of Holocaust and Redemption, today joins the Holocaust to the creation of the State of Israel. It links the secular and the theological, the Israeli and the diaspora communities of Israel."[38]

Since the establishment of the State of Israel in 1948, the area has experienced ongoing tension and violence. Efforts to establish States for both Israelis and Palestinians have been constantly thwarted. More than 1 million

Arabs are citizens of Israel, and today many Palestinians live on the Gaza Strip between Israel and the Mediterranean Sea, as well as on the Left Bank area.[39] Many think the "two state" framework is the only reasonable solution. The Palestinians struggle for a State of their own, though the United Nations did recognize Palestine in November 2012 as a nonmember observer state.

JEWISH SCRIPTURES AND TEXTS

There is a well-known story in rabbinic literature about a heathen testing two famous Jewish teachers, Shammai and Hillel. First, he went to Shammai and said: "Convert me on condition that you teach me the entire Torah while I am standing on one foot." Shammai was chagrined and drove the heathen away with a stick. Next, he went to Hillel, who converted him by saying, "That which is hateful to you, do not do to your neighbor. This is the entire Torah; the rest is commentary—go and learn it."

The story shows that "Torah" has many meanings: from the simple truth that love is central to life to "all Jewish study of Torah."[40] In the latter case, Torah would include the Bible, Mishnah, and Talmud, even though technically Torah refers to the first five books of the Hebrew Bible. Let's now take a look at these Jewish religious texts.

The Hebrew Bible – The Hebrew Bible, or Tanakh, is not so much a book as it is a library of many different kinds of literature: historical narratives, genealogies, poems, laws, prayers, plays, wise sayings, love songs, and prophecies. This library consists of books composed throughout many centuries by various authors in a number of different locations. At times, one narrative might involve the intricate weaving of a number of small pieces of traditions. Commonly, there are thought to be various sources, either authors or schools, at work in the Bible: the Yahwist (J), the Elohist (E), Deuteronomist (D), and the Priestly writer (P). These sources are dated from the eighth to the sixth centuries BCE. Even though this theory of authorship is challenged in some scholarly quarters, it is still commonly held to be accurate. Other sources from Near Eastern literature also seem to be present; they influenced such narratives as the creation stories and the story of the Garden of Eden, as well as the formulation of laws. In addition to all these sources, scholars have detected the work of redactors or those who have served as editors and even teachers in their own right within the texts.

The Jewish Bible is divided into three sections: Torah (the first five books and the most authoritative), Prophets (Nevi'im), and Sacred Writings (Ketuvim).

• *Torah*

The word Torah means teaching. The first of the five books of the Hebrew Bible, Genesis, teaches about pre-history: the time of creation, the development of humanity, and the calling of Abraham and his new nation of the Covenant. In the Book of Exodus, God's people are liberated from slavery in Egypt and, through their leader Moses, receive God's law from Mount Sinai. There are stories of their journeys, their rebellions against God and Moses, and their conquest of Canaan. Additional laws for rituals and everyday living are given in the final three books of the Torah: Leviticus, Numbers, and Deuteronomy.[41] In all, there are 613 commandments (mitzvoth) in the Torah: 365 prohibitions and 248 duties to be performed.

• *The Prophets*

The Prophets features the primary Hebrew prophets, who constantly chastise the people for their infidelity, repeated return to idols, lack of sincere devotion in their rituals, and neglect of the poor. Some prophets predict the famines, invasions, and exiles as punishments, while others assure God's people that God will be faithful to his Covenant and that they will be restored to their relationship with God.[42] The Prophets tells of the battles to win

Canaan, the beginnings of the monarchy under Saul, and the glory days of King David and his son Solomon, who built the Temple in Jerusalem. It chronicles the division of Judaism into two kingdoms, the conquest by the Assyrians and Babylonians, and the liberation of the Israelites by the Persians, who allowed the former Babylonian captives to return to Jerusalem.

• *Sacred Writings*

Sacred Writings is a treasure house of spiritual literature, which has nourished people for thousands of years. The Psalms are ancient prayers that cry out with many emotions:

Abandonment: "My God, my God, why have you forsaken me?" (Psalm 22:1).

Hope: "The Lord is my shepherd, I shall not want. He makes me lie down in green pastures; he leads me beside still waters. He restores my soul" (Psalm 23:1–3).

Love: "But I, through the abundance of thy steadfast love, will enter thy house" (Psalm 5:7).

Proverbs, a section in Sacred Writings, contains the astute insights of the sages: "When wisdom comes into your heart and knowledge is a delight to you, the prudence will be there to watch over you, and discernment be your guardian" (Proverbs 2:10–11).

"Pride comes first, disgrace comes after; with the humble is wisdom found" (Proverbs 11:2).

"He who trusts in riches will have his fall, the virtuous will flourish like the leaves" (Proverbs 11:28).

Sacred Writings also contains the gripping drama of Job, who questions why bad things happen to good people, as well as the salty cynicism of Ecclesiastes.[43] In this collection we also find the Song of Songs, stirring love poetry, which many think tells the story of God and his people; the Book of Ruth, which recounts the love story of Ruth and Boaz; the Lamentations, which mourn the destruction of Jerusalem and the Babylonian exiles; the Book of Esther, which describes Esther, the Jewish Queen of Persia, and her efforts to thwart a plot to kill her people; the Book of Daniel, which tells of the persecution of the Jews and the rise and fall of evil empires; and Chronicles, which retells the stories of the Israelites, at times differing in details from the earlier accounts.

The Mishnah – We saw earlier that during the rabbinic period following the destruction of Jerusalem and the Temple in 70 CE, work was begun on the Mishnah, the gathering of the oral tradition on the Torah. The Mishnah covers a wide number of topics, including Jewish law, accounts of legal disputes, stories, and interpretation of scriptural passages. It is the earliest text for teaching and may well have served as the curriculum used to educate new rabbis. The Mishnah is divided into six Orders and deals with many areas of Jewish life ranging from agriculture, family matters, food, Sabbath observance, and prayer. It contains halakhah, an important Jewish term that refers to matters of law, and aggadah, which deals with teachings, including stories, moral exhortations, and theological discussions.

Talmud – Ultimately, the Mishnah expanded into the Talmud and served as the core. Extensive Talmuds were developed both in Palestine and Babylon in the first half of the fifth century CE as the oral tradition increased among the rabbis through their vigorous debates and interpretations of Torah. Surrounding these Talmudic texts are the medieval commentaries of famous rabbis such as Shlomo Yitzhaki (1040–1105) and other commentaries called "Gemara." As mentioned earlier, there is also the greatest of the medieval codes, that of Moses Maimonides (1135–1204), who integrated Aristotelian philosophy into the discussion of Torah. Finally, the last comprehensive code—and one that influenced modern Jewish life—was produced by R. Joseph Caro (1488–1575).

Midrash – Midrash is a term that means interpretation or investigation. It was employed by the ancient rabbis who were attempting to draw meanings from the scriptures and apply them to everyday life. There is no single book called Midrash; rather, there are collections of Midrashim, which have been amassed and published in many volumes. Sermons preached in the synagogues and teachings of the sages were written down, edited, and published. Midrash came out of both halakhah (legal matters) and aggadah (teachings) in the sixteenth century. A treasured collection, with many down-to-earth stories, it is still used for study and as a resource for sermons.

Here is an example of Midrash:

A king wrote his wife a beautiful ketubah (marriage contract), promising her a wonderful living space and plenty of gold and jewels. The king then left her and went to a distant land for many years.

The wife's neighbors taunted her, saying that she had been deserted and should remarry. Though she was sad, she consoled herself by reading her ketubah. After many years, the king returned and said that he was amazed that she had waited for him all these years. She replied: "If it had not been for your wonderful ketubah, the neighbors would have won me over."

The lesson is that Israel is often invited to be loyal to other leaders and taunted that the Lord has deserted them. But when the Holy One comes and says he is amazed that they waited for him, they can say: "If it had not been for our study of the Torah which you gave us, the nations of the world would have led us astray."[44]

Another humorous anecdote shows how humans can be manipulators and even be so bold as to blame God for their misdeeds. This midrash is about God accusing Cain of killing his brother:

Cain says to God: "I killed him, true, but You created me with the evil urge in me. You killed him! You didn't accept my sacrifice and I was jealous."[45]

TWO CENTRAL BELIEFS OF JUDAISM

The history of the Jewish people has been twofold: the struggle to finally come to a belief in one God, and the ability to respond to the offer of a unique Covenant with that God. The belief in that one God and the commitment to be faithful to that Covenant, as well as the experience of God's fidelity to his people, has been the Jewish story. That same story has been central to the other two Abrahamic religions, Christianity and Islam. In the following section, we will discuss these two central beliefs of Judaism: God and Covenant.

God – The Hebrew Scriptures, as well as all the other teaching texts we have described, are solemn testaments to Israel's relationship with God. God's name is written as YHWH in the Bible and is not to be pronounced. Jews usually refer to God as Adonai. Non-Jews added the vowels "e" and "a" and wrote the name as Yahweh. Yahweh has often been interpreted as "I am who am," "He who is," or "He who brings all into being." It has been translated as "Lord" or "Lord God" or "Jehovah" in many English-language Bibles.

Central to Judaism is the belief that God is one, and this is repeated each day by the devout Jew in the sacred prayer, the Shema:

> *"Hear Oh Israel, the Lord, Our God, the Lord is One" (Deuteronomy 6:4). The one God is the creator of the universe and all that is in it, and the redeemer of all. Only this God is to be worshiped: "I am the Lord, and there is no other, besides me there is no God" (Isaiah 45:5).*

Many of the biblical narratives of the Israelites focus on the difficulty the people had in coming to a belief in one God. The story of Abraham and the culture of polytheism from which he emerged seem to be analogous to the multiple gods that the Semitic people had in their past and their consistent struggle to overcome idolatry. The account of the Israelites building a golden calf to worship while Moses is on Mount Sinai is an indication of how the chosen people periodically reverted to idolatry after their liberation from slavery. Enraged by their betrayal of the one God, Moses smashed the tablets of the Commandments given to him by God at Mount Sinai. The prophets in the Hebrew Bible also repeatedly warn their people against idolatry.

As late as the rabbinic period, concern about idolatry still existed. The rabbis classified as idolatry the Christian belief that Jesus is divine and warned their people against conversion. In the medieval period, Jewish scholars were also concerned that the Christian belief in Trinity, the concept of three persons in one God, undermined monotheism.

This one God in the Judaic tradition transcends all of reality: *"Have you not known? Have you not heard? Has it not been told you from the beginning? Have you not understood from the foundations of the earth? It is he who sits above the circle of the earth" (Isaiah 40:21–22).*

Isaiah goes on to say that this God is utterly beyond human understanding: *"For my thoughts are not your thoughts neither are your ways my ways" (Isaiah 55:8–9).*

In the history of Jewish thought there have been times, as in the case of the medieval Jewish philosophers, where reason was used to answer the perennial questions regarding God's nature and existence. In modern times some Jewish thinkers have approached the God question experientially and intuitively, grappling with the mystery by being open to the experience of God, rather than through rational speculation.[46] While the transcendence of God has prevailed in Judaism, there is also tradition regarding God's immanence. God is beyond all, but also within all. The Psalmist prays: "Whither shall I go from your Spirit? Or whither shall I flee from your presence? If I ascend into heaven, you are there!…If I take wings in the morning and dwell in the uttermost parts of the sea, even there your hand shall lead me" (Psalms 139:7–10). The rabbis were fond of comparing this all-embracing presence of God to the light that shines on every part of creation.

God's omnipotence was firmly held in biblical times. The Lord said to Abraham that the barren Sarah would bear a child: "Is anything too hard for the Lord?"(Genesis 18:13–14). And God told Jeremiah: "Behold I am the Lord the God of all flesh: is anything too hard for me?" (Jeremiah 32:27). This omnipotence was questioned by some of the medieval Jewish philosophers, and in the modern period some perceived "God's absence" in the Holocaust and raised serious questions about God's omnipotence. Richard Rubenstein wrote: "We stand in a cold, silent, unfeeling cosmos, unaided by any purposeful power beyond our own resources. After Auschwitz, what else can a Jew say about God?"[47] Rabbi Kushner addressed this question in his best-selling book *Why Bad Things Happen to Good People*. The rabbi had lost his only son to progeria, rapid-aging disease, and was heartbroken. At first he blamed himself, thinking that perhaps it was because of his own failings that the boy died. There had to be a reason that the all-powerful God sent this disease to his son. Then the rabbi began to question his belief in the omnipotence of God. Maybe God has no power over ill fortune or nature itself. Maybe God's power existed in his love and compassion during the trials of life, rather than in controlling all things.

God's omniscience was also part of the biblical tradition. Psalm 33 proclaims that "the Lord looks down from Heaven, He sees all the children….He who fashions the heart of them all observes all their deeds" (Psalms 33:13–15). The Psalmist also believes: "You know when I sit and when I rise up; you know my thoughts from afar (Psalms 139:2–3). The Jewish philosophers found it difficult to reconcile God's omniscience with free will. Once again, in modern times, the horrendous experience of the Holocaust raised serious questions about God knowing all things and yet not coming to his people's aid.

The Jewish way is often to question and even challenge God when tragic situations arise. There is a story in the Talmud telling how Moses challenged God about the horrible death a rabbi suffered at the hands of the Romans. Moses asked God: "Master of the Universe, this is Torah? And this is the reward?" God answered: "Silence! That too has occurred to Me." (According to the Talmud, even God sometimes doesn't understand.)[48]

Covenant – Some scholars think that the Sinai Covenant was modeled after the treaties that ancient kings made with their people. These treaties were read publicly across the realm several times a year and are similar to the biblical covenant in that they begin with the title of the king and establish a relationship with the subjects, show the historical basis for the agreement, set down the obligations of the people, and then set down blessings as well as threats to them, depending on their fidelity to the law.[49] The covenant is sealed with people making a pledge to be faithful.

In the case of the Sinai Covenant, the king is God; the historical basis is that he brought the people out of Egypt; the obligations are the Commandments; and the blessings are steadfast love and protection for fidelity. Defeats, exiles, and other punishments result from breaking the obligations. (The conditions and warnings of God's wrath are constant in the prophetic tradition.) The Covenant is then accepted by the Israelites, who pledge to abide by the Covenant obligations and thereby become the people of Yahweh.

The biblical Covenant, then, is a solemn agreement of fidelity between God and God's people. Biblical scholars have observed two kinds of covenant in the Bible: the unconditional and the conditional. Examples of unconditional covenants are the one made to Abraham, promising the land of Canaan to his descendants as an "everlasting possession" (Genesis 17:8), and the promise of an "everlasting covenant" with the yet unborn Isaac (Genesis 17:19). A similar covenant was made with the house of David (2 Samuel 7).

Conditional covenants differ in that God promises blessing if Israel keeps the law and performs God's will, but also promises curses and political and natural disasters if Israel is unfaithful. This type of covenant is associated with that made with Noah as well as with the covenants made in the Exodus and Sinai traditions, as described above.

The Covenant is the basis for the Jewish belief that God's love is eternal and that God will always be forgiving of his people, even when they have broken their part of the Covenant. It is also the basis for the Jewish belief that they are "the chosen people of God." Being "chosen" has a number of interpretations by the Jewish people. Rabbinic Judaism believed that God chose Jews from all peoples and that this election was due to their acceptance of the Torah. Today, many Jews point out that being chosen does not mean that they are superior or are people deserving service. As one scholar puts it: "Judaism impels individuals to serve others. In the view of the prophets, the Jewish nation is to be dedicated to service." [50]

CELEBRATIONS AND FEASTS

Devout Jews pray throughout the day and the year. Jacob Neusner says that "to be a Jew in the classical tradition, one lives his or her life constantly aware of the presence of God and always ready to praise and bless God."[51] It is believed that such devotion shapes both individuals and communities into caring and loving people. Jews are a people who celebrate time and events as sacred. Abraham Heschel (1907–1972), one of the great rabbis of our time, wrote: "Judaism is a religion of time, aiming at the sanctification of time…. Judaism teaches us to be attached to the holiness in time, to be attached to sacred events, how to consecrate sanctuaries…."[52] Heschel points out that though the Jewish temples and synagogues have been destroyed in the past, no one can destroy the sanctuaries of Jewish Sabbaths, feasts, and rituals.

Celebration of the Sabbath (Shabbat) is central in Judaism. The Sabbath begins at sunset Friday and lasts until dusk on Saturday. It is celebrated in honor of God's day of rest after creation was complete. For Jews all around the world, the **Sabbath** is a sacred time to enjoy family, prayer, and a peaceful retreat from school and work. First, the Shabbat candles are lit to welcome the Sabbath, and prayers are recited for domestic harmony. The challah bread is blessed before the meal, the kiddush blessing is recited, and wine is drunk with special wine cups to demonstrate happiness. Saturday morning, families go to synagogue where there will be special readings of the Torah and prayers. After the service, they return home and enjoy the day together in the presence of each other and their Creator. Prayers and blessings are said before each festive meal, and only kosher food is served (i.e., meat specially butchered and food specially prepared by authorized distributors).

At the turn of the year, Jews celebrate the most solemn of the feasts, the High Holy Days: **Rosh Hashanah**, which is the Jewish New Year, and **Yom Kippur**, the Day of Atonement. These ten days are a time for self-examination and repentance. On Rosh Hashanah the family attends synagogue to celebrate the birthday of the world, and they pray: "Our God and God of our Fathers, rule over the whole world in Your honor…" The shofar, or ram's horn, is sounded in the synagogue, and the people are reminded of God's judgment on their good or bad deeds. On the most solemn Day of Atonement, a day of fasting and prayer, the family gathers in the synagogue for a day of remembrance of their sins and confidently asks for God's love, compassion, and forgiveness.

Five days after Yom Kippur, the Feast of Tabernacles, known as **Sukkot,** is celebrated. Prayers of thanks are recited for the crops being harvested, for their food, and for the rains that are to come. Eating outdoors in a simple hut built behind their house, they celebrate the time their people spent in the desert and enjoy the wonders of nature, God's gift of creation, and liberation.

In the spring, Jews observe the great feast of **Passover**, a celebration of the most important event in Jewish history known as the Exodus, God's liberation of the Hebrews from slavery in Egypt. Its name, Passover, refers to the stories of the plagues God inflicted upon the Egyptians, including one that would kill the firstborn sons of Egypt. The Hebrews were told to mark their doorposts with lamb's blood so they would be passed over when the firstborn males of Egypt were killed. On this most extraordinary night of the year a special **Seder** meal is offered at home, with solemn readings about the Exodus. Bitter herbs, salt water, and a mixture of fruit and wine are served to symbolize the difficult times of slavery in Egypt. Unleavened bread is also eaten to commemorate that the Hebrews left Egypt in haste and did not have time to put yeast in their bread. Jews recall what it means to be God's chosen people and to count on his love and freeing power in their lives as they eat these symbolic foods.

Seven weeks after Passover, the feast of Pentecost (**Shavuot**) is celebrated, another feast of harvest and a commemoration of the giving of the Torah at Mount Sinai.

》 *Learn more about both the celebration and the origins of Passover as you watch Passover Parts 1–2 :* http://www.youtube.com/watch?v=0d49a8tceis&featur e=related and http://www.youtube.com/watch?v=u2QZiJp9_3o&feature=related.

Another important festivity is **Hanukkah**, which celebrates the Maccabean victory over the Syrians, the rededication of the Temple in 164 BCE, and the lighting of the menorah or Hanukkah lamp. At this time, religious freedom is highlighted. The family lights one candle each night for eight nights to commemorate the event, and on the last evening there is a special meal. Since this is the traditional holiday season, gifts are often exchanged.[53]

Purim is a minor feast and commemorates Esther, the Persian queen, who saved her people from destruction from evil. Children wear costumes and masks and have fun acting out the story of Esther for their family and relatives. Special pastries called "Haman's ears" are served.

Weddings are always joyous occasions for Jews. Before the ceremony, the groom signs the ketubah, a document that promises he will pay a certain sum to his bride if he should divorce her. This agreement provides security for the wife and helps to assure that the husband will give serious thought before considering a divorce. The couple joins the rabbi under a canopy (chuppah) where blessings are recited, and the bride and groom sip from a cup of wine. The groom then performs the betrothal ceremony (kiddushin) by giving the bride the ring while proclaiming: "Behold, you are consecrated to me with this ring in accordance with the law of Moses and Israel." The ketubah, which the groom signed earlier, is then read and the wine is again sipped. Then the groom crushes the glass with his foot as a remembrance of the destruction of Jerusalem.

Funerals – Funeral services for the dead differ among Jews. Usually there are some biblical and liturgical verses recited, a eulogy is given, and special memorial prayers and the Kaddish prayers are recited. Generally a mourning period is kept, sometimes for seven days. Traditional Jews bury the deceased as soon as possible, with no embalming, and permit only a plain wooden coffin. A marker is placed on the grave. Reform Jews allow embalming and cremation.

>> *Visit the website Becomingjewish.org and select from the many 10-minute topics to learn more about Jewish celebrations and feasts.*

CONTEMPORARY JUDAISM: CONNECTING JUDAISM TO WORLD ISSUES

Today about 80 percent of the Jewish population live in Israel and the United States, with nearly 6 million Jews in each country. There are more than 1 million Jews in the European Union and relatively small numbers in Africa, Eastern Europe, and South America. A large percentage of Jews identify themselves as "secular," though they may go to synagogue from time to time.

In Europe, Orthodox is the largest movement, but in the United States the Reform movement has the largest membership, and the Conservative movement has the second largest. Both Reform and Conservative have recently made some headway in Israel, which is predominantly Orthodox. No matter their differences, there is solidarity among Jews, all of whom share a common history and culture. As it says in the Talmud: "All Jews are responsible for one another."

Although demographically a relatively small group, Jews have won a large number of Nobel Prizes and have achieved fame in a number of areas. To mention a few: Boris Pasternak (d. 1960), Saul Bellow (d. 2005), and Elie Wiesel in literature; Yehudi Menuhin (d. 1999) and Itzhak Perlman in music; Albert Einstein in science; Henry Kissinger and Yitzhak

Rabin in politics. There are currently three Jewish Justices on the United States Supreme Court: Stephen Breyer, Ruth Bader Ginsburg, and Elena Kagan. Alan Greenspan and Ben Bernanke, who have served as chairmen of the Federal Reserve, are accomplished in finance.

Jews in the United States – The United States has one of the largest and possibly most vibrant and diverse communities of Jews in the world. Thirty-nine percent of practicing Jews belong to the Reform Movement, thirty-three percent are Conservative, twenty-one percent are Orthodox, and three percent are Reconstructionists.

In the 1960s and 1970s there was serious concern about anti-Semitism in the United States, especially from Neo-Nazi groups and from some in the evangelical Religious Right who maintained that the United States was a "Christian Country."[54]

One of the groups that work to combat anti-Semitism and protect Jewish civil rights in the United States and abroad is The Anti-Defamation League. The League, formed in 1913, works to stop the defamation of Jewish people by appealing to reason, to conscience, and if necessary, by appealing to law. As a civil rights/human relations agency, the League works to end all types of bigotry and eliminate discrimination and ridicule directed toward any sect or group.[55]

There has also been some concern that Jews were losing their identity in the United States, especially through intermarriage and the many "secular Jews" who do not practice their religion. To deal with this, the Reform Movement has taken steps to ask Gentile partners to convert to Judaism and proposed that birth from either father or mother qualified one to be Jewish. Traditionally, one was considered Jewish only if born from a Jewish mother.

"Continuity" has become a watchword for many Jews in the United States, and there is growing commitment to the study of Torah, expansive education to youth in the traditions of Judaism, and observance of Jewish practices and celebrations. Especially among the Reform congregations, there is now openness to converts from other ethnic groups such as Latinos. These congregations also are accepting members, couples, and rabbinic candidates who are gay or lesbian.[56]

As in other religions, many American Jews today are often more interested in spirituality than they are in religion. Many are less concerned about laws than they are about experiencing God personally and in community, preferring to look to the Jewish tradition more as a guide for their ongoing search for meaning and purpose. And, as we will see in the next section, more Jews are connecting their faith with social issues.

Jewish Involvement in Social Issues – Today, Jews have become strongly identified with the cultures in which they live and work. Consequently, they have also been affected and galvanized by pressing contemporary issues and have become more active in leadership. Here we have selected only three of the many issues: ecology, peace, and the women's movement.

Obviously, worldwide, Jews are experiencing the environmental crisis and have become proactive in the movements to lessen the impact of their environmental footprint. Ongoing tensions and violence in and around Israel and the many wars in surrounding countries have moved Judaism to renew its age-old traditions on peace. Lastly, the global women's movement has raised the awareness of inequality in Jewish women's minds and has moved them to work for fairer treatment.

• *Jewish Involvement in Ecology*

Jews have only recently begun to connect their tradition with ecological concerns, and some are searching their tradition for insights that apply to ecology. Jeremy Benstein, a Jewish environmentalist, points out that like the rabbis of old, Jews need to do new midrash (interpretation), where "not only are we making claims on the text in our creative interpretations, but the texts are also making claims on us."[57] Whereas in the past, Torah assisted in explaining their relation to God, now Torah is helping the

Jewish people to relate to God's creation in new ways. Diversity among Jews, ranging from secular to Orthodox, makes this new examination of Torah that much more challenging.

There is a flourishing environmental movement in Israel. Many Israelis realize they have been so preoccupied with development, industrialization, transportation, and security that sufficient attention has not been given to the resulting air pollution and health hazards, including a marked increase in breast cancer and asthma. A growing population of nearly 8 million, four times what it was fifty years ago, is now aware that serious steps must be taken to improve its environment and to lessen its carbon footprint. It is also hoped that environmental concerns will be a way for both Israelis and Arabs to establish common ground amid their many conflicts.[58]

These connections between Torah and ecology are being made worldwide. In 2009, Jewish rabbis, environmentalists, economists, and educators from around the world gathered in Jerusalem to draft the "Seven-Year Plan for the Jewish People on Climate Change and Sustainability." This document will ultimately be joined with the similar plans from other world religions. In the following section, we examine some of the beliefs that Jews have linked to ecology.

The Earth is the Lord's – A central belief in Judaism is that God is the creator of all things and that ultimately creation belongs to God. The Psalmist writes: "The earth is the Lord's and all that is in it, the world, and those who live in it" (Psalm 24:1). "The heavens are yours, earth is also yours; the world and all that is in it—you have founded them" (Psalm 89:11).

Caretakers of the Earth – In the Jewish tradition, humans have been given the privilege "to cultivate and care for" the earth (Genesis 2:15). In *Genesis 1*, humans are told to fill the earth, subdue it, and take dominion over all living things. More and more, Jews have come to realize that this role of stewardship has often been seriously neglected.

In the daily morning service, Jews recite Psalm 148: "Praise God, sun and moon; praise God, all you shining stars!" They join the heavens and all living creatures to praise their Creator. Many Jews now view such prayers as a new calling to see their interconnection with nature, its beauty and fragility, and be more resolved to sustain the earth and its resources.

Connecting Torah with ecology has given new insights to traditional celebrations. Sabbath has always been a celebration of sacred time and sacred space. With a renewed interest in their earth, many are incorporating into their Sabbath observance an appreciation of the beauty around them and a stronger resolve to sustain that beauty. The Jewish calendar and celebrations are based on the positioning of the sun and moon, the harvests, the stages of life, and sacred times of the year. Many Jews now celebrate these feasts with a new vision. "It is a vision of life regenerating itself, as part of the cosmic process, akin to nature and its renewal in the cycle of the seasons and the years, which finds expression in the calendar and our marking of time."[59]

The People of God – The Jewish people have always understood that they have been specially chosen by God to live within a Covenant that requires them to keep God's law. Many Jews see themselves as a light unto all the people of the world. Central to that Law is to love God and to love neighbor as one loves self. Now there is a new awareness that this responsibility includes helping the world care for creation.

The Jews have a strong sense of being a people. Ellen

Bernstein, a pioneer thinker who is recognized as helping to define Jewish environmentalism, points out that the early rabbis understood "that humanity was inherently arrogant, and arrogance could lead people to destroy their neighbors, the natural world and themselves."[60] Now Bernstein believes that this communal dimension is being discovered in a new way, with a stronger sense of how harming nature harms the self and others.

There is an ancient story told by the rabbis about two men who were out on the water in a rowboat, when suddenly one of them took a saw and began to cut a hole under his feet. When asked why, he said it was his right to do whatever he wanted with the place that belonged to him. The other man reminded him that they were in the rowboat together and that the hole would sink them both.[61] The moral of the story is obvious—we are all in this vessel called earth together and are responsible for each other.

Jews in Action

Judith Helfand – Judith Helfand is a young American Jewish activist, educator, and filmmaker, who is passionate about the problems of chemical exposure, corporate irresponsibility, and environmental injustice. Her film *The Uprising of '34* won the Sundance Award, while *Blue Vinyl* was nominated for an Emmy. Judith says that in her films she likes to tell stories that both entertain and challenge people to be more socially conscious. She has received "America's finest living artists" award, among many others.

>> *Watch the trailer to Judith Helfand's film Everything's Cool*
http://www.youtube.com/watch?v=I_NqJ59Y3XA
to learn more about her commitment to spreading the word about environmental threats.

Alon Tal – Alon Tal is a professor at Ben-Gurion University of the Negev in Israel and holds a law degree from Hebrew University as well as a doctorate from the Harvard School of Public Health. He is an award-winning activist in protecting the environment of Israel. Tal was the founder of the **Israel Union for Environmental Defense** and the Arava Institute for Environmental Studies. He has contributed valuable research into water management and has been active in working with Palestinians on the impact of hazardous material, air quality, nature preservation, and the disposal of solid waste. Professor Tal has represented his country at the United Nations on desertification and oversees the preservation of forests, reservoirs, and rivers in his country. He is founder and co-chair of an environmental and social political party, the Israel Green Movement.

>> *See Prof Alon Tal* http://www.youtube.com/watch?v=nHICEBGWrcE&feature =PlayList&p=90F839F1D038539D&playnext_from=PL&index=0 **discuss the need for an Israeli political party to deal with the challenges of a changing environment.**

• *Judaism and Peace*

Peace or *shalom* is held up as an ideal for the righteous and liberated people of Israel. In an often-quoted passage, the prophet Isaiah projects his vision of Israel at peace:

"He shall judge between the nations, and shall decide for many peoples; and they shall beat their swords into plowshares, and their spears into pruning hooks; nation shall not lift up sword against nation, neither shall they learn war any more" (Isaiah 2:4).

The Psalmist cries out for the blessing of peace: "May the LORD give strength to his people! May the LORD bless his people with peace!" (Psalm 29:11). The common Jewish greeting is Shalom aleichem, which means "Peace be with you." Many of the Jewish prayers include petitions for peace.

While Judaism does not have a tradition of pacifism, it does see peace as the highest good. Lawrence Schiffman, a Jewish scholar, points out that Judaism condemns revenge and unprovoked aggression, but it allows for preservation.[62] In Judaism, peace is with God, self, and others, but this does not necessarily imply a political peace. Torah requires warning enemies that if they continue to be adversarial, they will be attacked. Torah also orders that in victory, there should be limited damage to the environment (Deuteronomy 20:9–12).

The Holocaust elicited many examples of the Jewish approach to violence. Many accepted their torture and abuse with dignity, determined not to sink to the level of their oppressors. There is a story about a rabbi and another Jewish man being forced to watch a fellow inmate being beaten. The man asked the rabbi why Jews are so often persecuted. He answered: "In this camp, where you can only be a victim or a perpetrator, we should be proud we are the victims."[63] The establishment of Israel as a nation was followed by a series of wars with its Arab neighbors. For many years it has been locked in a struggle with Palestinians. There are debates on all sides as to how these conflicts can be settled. Both in Israel and the United States there are peace groups such as **Peace Now** and **American Jews for a Just Peace** that work for a peaceful settlement.

Generally, Jewish people hold to the just war theory. Some believe that the early biblical wars were just because they were ordered by God, but others take a more critical view of Scripture and interpret these war stories as more symbolic than historical. The just war theory provides a number of conditions: fighting is a last resort; it is authorized by legitimate authority; it is carried out in self-defense; there is reasonable hope for success; the goal is peace; the violence is proportionate to the offense; and the killing of non-combatants is avoided.

Jewish thinkers debate the use of nuclear weapons. Some hold that such weapons are moral if used as a deterrent, but not otherwise because of the massive loss of innocent life. Others maintain that nuclear weapons may be used for self-defense. Some oppose nuclear weapons on the grounds that all weapons are eventually used at some time, and that nuclear weapons, in fact, provoke war.

In today's world, where terrorism is a serious threat, Jews are opposed to suicide bombings and to the taking of innocent life by terrorists. The Jewish tradition values the sacredness of human life and love of God, self, and neighbor. Forgiveness and mercy are cherished values. All these ideals are applicable to their commitment to peace.

The Sacredness of Human Life

Genesis 1 tells a story of creation wherein humans are created on the sixth and final day. The Creator God gathers his heavenly consort and says: "Let us make humankind in our image" (Genesis 1:26). Dan Cohn-Sherbok, a Reform rabbi and Jewish scholar, points to the ancient Jewish tradition that because people have been created in the image and likeness of God, they hold a responsibility to imitate God's goodness and love.[64] In the Jewish tradition, the cosmic Creator of "good things" blessed his godlike creatures and gave them dominion to care for all the living things on earth. They are in solemn covenant with God and stand as a holy people, who are representative of God's love and saving power to the nations. Humans are empowered "to multiply," to produce other "images" of God (children), and to treat them with love and respect.

In *Genesis 2*, the very life of the human comes from the breath of God; life finds its source in the divine and is therefore to be cherished. In the Jewish tradition people are created to be companions with God and faithful followers of God's law. In *Psalm 8:5*, the human being is described as crowned with glory and "a little lower than God." Although the Jewish tradition also recognizes that humans are sinners and can be unfaithful to their God, humans are constantly forgiven and called back to their true and dignified selves, called to return to peace and justice.

One of the prayers for Yom Kippur sums this up: "Our God and God of our fathers, hear our prayer, do not ignore our plea. We are neither so brazen nor so arrogant to claim that we are righteous, without sin, for indeed we have sinned....May it therefore be your will, Lord our God and God of our fathers, to forgive us all our sins...."[65]

Love of God, Self, and Neighbor

Jewish spirituality exists within a triangle of love involving God, neighbor, and self. The Torah commands: "You shall love the Lord your God with all your heart and with all your soul and with all your might" (Deuteronomy 6:5). One of the commandments in Leviticus is: "Love your neighbor as yourself" (Leviticus 18:18). In the Talmud, there are numerous discussions of how to do *mitzvah*, or good deeds for others, in order to add to one's status before God. This involves a commitment to respect and honor one's family as well as the members of one's community. The love of God, neighbor, and self are all essential to salvation. To neglect one element is to neglect the other two. Likewise, to be faithful to one is to be faithful to all. In the Mishnah, the great teacher Hillel says: "Be one of Aaron's disciples, loving peace and pursuing peace, loving mankind and bringing them near to the Torah."[66]

Jews in Action

Elie Wiesel – Elie Wiesel (b. 1928) was born in Sighet, which is now part of Romania. When he was fifteen, he and his family were deported to Auschwitz. His mother and younger sister died there, but his two older sisters survived. Elie and his father were later sent to Buchenwald, where Elie's father died. Upon his liberation from the camp, Elie traveled to Paris and became a journalist. He wrote an account of his experience in the camps in *Night*, which became a bestseller and was translated into thirty languages.

Wiesel has been a tireless advocate for oppressed people and refugees around the world, no matter their nationality. He has also been an ardent supporter of Israel. In 1980 he became the founding chairman of the **United States Holocaust Memorial Council**. Wiesel has won numerous awards, including the Presidential Medal of Freedom and the Nobel Peace Prize in 1986.

Father Bruno Hussar – An organization that Elie Wiesel champions is called Neve Shalom/Wahat al-Salam. Founded by Catholic convert Father Bruno Hussar in 1972, the vision was to build understanding among Christians, Jews, and Muslims. Hussar dreamed of a village where all would live together in peace, where their diversities would be a source of enrichment rather than division. In 1972 a few Jewish and Arab families joined him, and the community was born. The soul of the village, the School of Peace, has become a magnet for peace-seeking Jews, Palestinians, and others. More than twenty thousand young people have been trained in ways of peace. They try to spread their message of tolerance throughout the world.[67]

Michel Warschawski – Michel Warschawski is a Polish/French Jew, who grew up in France with Orthodox parents. His father fought in the resistance movement during the Nazi occupation. As a young man, Michel went to Jerusalem to study Talmud. He enjoyed "living on the borders" where languages, food, and beliefs were exchanged. He soon became a political activist involved in radical politics. Michel's passion was creating inspired alliances among Jews, Muslims, Christians, and atheists. As the borders hardened, Michel took on the code name of Mikado and became a leftist movement leader. He was soon targeted by Shin Bet, Israel's intelligence agency, arrested, and jailed for twenty days. When he was released he began writing about the psychological pressures he received in jail and about the severe treatment that was given Palestinians. His book *On the Border* is an interesting account of his experiences.

Hagit Ra'anan – Hagit Ra'anan is an Israeli who lives in Tel Aviv. She has been working for peace for many years. Her grandfather took the family from Poland to save them from the Holocaust. Her parents were part of the underground movement Etzel fighting for Israeli independence. By age thirty-two she had experienced six wars, served in the army herself, and had lost many of her friends in battle.

In the Lebanon War in 1982, Hagit lost her beloved husband, who was killed by a Palestinian. She then miscarried their child. These events plunged her into a depression, but a healing and meditation group helped her become whole again. Hagit says she carries no guilt or blame: "I'm not a victim and don't believe any of us are victims. I have come to the conclusion that staying in a victim position brings no good. I have to do something and move on." She decided to move in positive directions and help others do the same.

After the assassination of the Israeli Prime Minister Yitzhak Rabin in 1995, Hagit realized that not only individuals but also nations need healing. Before the Intifada (the uprisings of Palestinians against Israel starting in 1987) she would bring Israelis to the refugee camps so that they could get to know Palestinians. She believes fear breeds violence and getting to know people takes away fear. Terrorist attacks in Israel sadden her because she knows that Israelis are killed or injured. At the same time she knows there will be revenge and her "other children" will suffer. Hagit says she is not political. After the loss of her own baby, she has tried to be mother to all the children she encounters. She says: "There are many children in the Gaza Strip and in many of the cities and refugee camps in the West Bank who call me *Yamma* (Mommy) or *Sitta* (Grandma). At this time, my main service is humanitarian assistance: helping people, visiting injured Palestinians at the Israeli hospitals, bringing clothes, food, and medicine, whatever I can possibly do to ease the situation."

>> **See Jerusalem Peacemakers:**
Hagit Ra'anan: Jerusalempeacemakers2008.jerusalempeacemakers.org/hagit/home.html

• *Jewish Involvement in the Women's Movement*

Torah is both positive and negative toward women. The two creation stories give the genders a certain equality by saying both were created together (*Genesis 1*) or that both came from the same bone (*Genesis 2*). The Covenant is made with both women and men, even though only men receive its mark of circumcision.

At the same time, in the biblical world women had few rights in marriage or in court, did not study Torah, and after being "purchased" for marriage, spent their days keeping the home and raising children. The Bible evolved within the patriarchal structures that prevailed at the time, and even though women were seen as equal in the eyes of the Creator and were welcomed in Temple, their place was in the home. If they did not measure up, they could be given a bill of divorce. Men held the positions of authority in government, court, and religion. Men were to teach their trades to their sons, not their daughters, and thanked God in their morning prayers that they were not born as women.

In the rabbinic period that followed the destruction of Jerusalem in 70 CE the rabbis improved the woman's lot in marriage. Women were no longer purchased in marriage, and their consent was required. The marriage contract (*ketubah*) gave wives the right to some funds if her husband died or divorced her and also allowed her rights to the estate of a deceased husband. Wives now had the rights to participate in business, to have medical care when ill, and to receive a "writ of divorce." Some rabbis annulled marriages, thus allowing women to remarry. There seems to be a presumption in the ancient Jewish tradition that women are seducers, because the *Mishnah* forbids men to be alone with a woman other than his wife.[68] Women were also deemed ritually "unclean" by virtue of their menstruation and giving birth.

Once religious practice shifted from Temple to synagogue, women were not required to attend, but were still allowed leadership in the home ceremonies such as Sabbath service. Ultimately, this led to the home achieving more religious importance than the synagogue.[69] Without attendance at religious services or the possibility to study Torah, women had to nurture a domestic piety.

Without much access to wealth or education, many women remained illiterate and subservient. And even though the rabbis granted that women were equal, they also insisted that men and women were not the same and should complement each other with their different roles: women as mothers and homemakers, men as providers and leaders in public positions.

The Women's Movement – In the last fifty years women have made their voices heard and effected significant changes in Judaism with regard to women's rights. Women have challenged the patriarchal structures and gender roles in Judaism. They have pointed out that Torah, Talmud, and other religious texts have been written by men; therefore, the texts lack feminine experience and perspective. They note that thousands of rabbis throughout the centuries compiled the Talmud, and only several times made reference to women! Women scholars have pointed out that God has been portrayed as a man, a warrior, and a judge, while little attention has been given to the feminine images of God in the scriptures. They point out that the language in their biblical translations and prayers focused on "men" and were not inclusive of women. In order to correct these oversights, many Jewish women expect a place in biblical, Torah, and liturgical studies so that they can provide a feminine perspective.

Women have also sought equality in religious services. Some have asked to be ordained as cantors and rabbis; others wanted to wear the prayer shawl and to be called to the pulpit to recite blessings and read Torah. Some women also want more recognition in Jewish life, such as admission into the rabbinic schools, to be part of the synagogues' decision-making bodies, and to have access to leadership roles in their religious communities. In 1974 the Jewish Feminist Organization bluntly stated that women wanted "nothing else than the full, direct, and equal participation of women at all levels of Jewish life."[70]

Some Jewish women want to transform Torah into a renewed tradition that acknowledges the dignity and equality of women. This would require new commentaries on Torah, which recognize its patriarchal limitations and open the way for building a renewed community that emphasizes human dignity rather than the duality of male and female. Judith Plaskow, a leading Jewish scholar, calls this "the recovery of the fullness of Torah."[71]

Considering the many centuries of patriarchal traditions, the changes came quickly, but with differing responses among the Jewish movements. The Orthodox, which views Torah as written by God, had the most difficulty making changes. The Reform and Conservatives, which accept the human authorship of Torah, decided on changes with more ease.

The Reform movement made significant changes early on. The first woman Reform rabbi, Sally Priesand, was ordained in 1972. Since then, many others have been ordained. In Europe there had been some earlier examples of women rabbis. Hannah Rochel Verbermacher (1815–1892), a Ukrainian, served as a rabbi, taught Torah in her community of Ludmir and later in Jerusalem, and had a large following of disciples. In 1935 Regina Jonas was ordained a rabbi in Germany. She served as a rabbi during the Nazi regime, working as a scholar, teacher, chaplain at a hospital, and champion of women's rights, including ordination. Regina was a Nazi resister and died in Auschwitz in 1944.[72] The Reform movement also has female cantors, and women participate as equals in all services. Bat mitzvahs are held for girls and laywomen. Women are now able to study and teach Torah and serve in the Jewish congregations in positions of authority. The language of the prayers, liturgical texts, and religious textbooks is now inclusive.

Changes in the Conservative movement evolved more slowly. This movement also recognizes the human authorship of Torah and is open to change, but at the same time is very careful not to let secular values or fads dictate alterations in the tradition. Many believe that the Torah tradition gave women more dignity than secular cultures and that Judaism lost much in conceding to secular values.

After much serious debate, the Conservative movement in 1980 voted in favor of a resolution to ordain women rabbis, and women were gradually admitted into the rabbinic programs. Women have also been admitted as cantors. They now have full participation in the religious services and read Torah. Many laywomen take advantage of Torah study and form study and prayer groups of their own.[73]

Orthodox Judaism has been resistant to changes in the role of women in their congregations. Some have established parallel rabbinic study programs, and women can serve as Rabbinic Assistants. Any hope of ordination, however, seems to be far in the future. However, women have been given more opportunities to learn Torah and *halakhah* (the ethical teachings). Some Orthodox congregations have introduced a modified bat mitzvah for girls. Many Orthodox women have formed their own prayer groups and even conduct their own services. Women are also being allowed to serve as voting members and leaders of congregational committees.

Some Orthodox women have joined the Women of the Wall, who pray at the Western Wall in Jerusalem and demonstrate their displeasure with those who resist change with regard to the role of Jewish women in the faith. They have had to face opposition and sometimes even violence from ultra-Orthodox groups. They have been fired upon with tear gas, physically assaulted, and threatened with years of imprisonment if they read publicly from the Torah scroll, blow the ram's horn, or wear prayer shawls.

Jewish women have turned to a number of teachings in the

Torah to achieve their liberation. Among these are: a God beyond gender, the fullness of Torah, and equality of all humans.

A God Beyond Gender – Though the male pronoun is used in Torah, the God of the Israelites is nameless and transcends gender and imaging. As God says to Moses: "I am Who Am." Although the common imagery for God is male in Torah, women scholars are quick to point out that feminine imagery is also used, and God is described as mother in *Isaiah*: "Can a woman forget her nursing child, or show no compassion for the child of her womb?" (Isaiah 49:15–16). "As a mother comforts her child, so I will comfort you" (Isaiah 66:12–13). *Hosea* also speaks of the Mother God: "I was to them like those who lift infants to their cheeks. I bent down to them and fed them" (Hosea 11:4). And in the *Book of Wisdom* there is the magnificent image of Lady Wisdom who partners with God in creating.

Women scholars point out that the images of God as king of the universe and Lord only serve to confirm hierarchy among people and between the sexes. Such images also seem to call for passive obedience. Elliot Dorff, a Jewish scholar, points out that the traditional overemphasis on God as transcendent and impersonal does not appeal to many Jewish women. He points to the deep personal relations with God that Jews have in their tradition and urges approaches to God that are more earthy, intimate, and deeply rooted in the human experience.[74] Female scholars promote images of God that will empower them to courageously give themselves to improve their world. Some women in their own prayer groups find rest and peace in Shechinah, the gentle and loving presence of the divine feminine.[75]

Some Jewish women want to participate in this new reimaging of God. As Judith Plaskow writes: "Only when those, who have had the power of naming stolen from us, find our voices and begin to speak, will Judaism become a religion that includes all Jews—will it truly be a Judaism of women and men."[76]

• *The Fullness of Torah*

Since Torah, Talmud, and other texts of Judaism were written by men, many Jewish women think it important in our era to look at these writings from a feminine perspective. Examples of patriarchy, hierarchy, and other elements that many women believe to be more cultural than revelatory may be, therefore, construed as subject to change. At the same time, women want to reinforce beliefs and themes that are supportive of women, but which have been passed over through the centuries. They also want to reexamine the stories of biblical figures such as Eve, Sarah, and Ruth and offer new lessons from the feminine viewpoint.

Women also bring new eyes to the laws of Judaism, to see which of these laws are unfair or unjust to women's rights. They expose cultural biases from ancient times that are no longer appropriate.[77] Women listen attentively to the voices and experiences of women as they live the rich spirituality of Judaism: women who survived the horrors of the Holocaust, as well as women of Africa, Asia, the Middle East, and all parts of the globe. Such women can bring many new perspectives to Torah and Talmud.[78]

Earlier we discussed how both creation narratives reflect an early gender equality in Judaism. In contrast, the Jewish tradition often gives assigned and complementary roles to woman and men: the women in the home and men in public life. This social structure has often led to the exclusion of women from education and participation in public ritual, as well as limiting women's rights in the court system. The patriarchal and hierarchical structures within Jewish society throughout the centuries have, for the most part, demoted women to an inferior position.

Jews in Action

Susannah Heschel – Susannah Heschel is the daughter of the famous Rabbi Abraham Heschel. She is professor of Jewish Studies at Dartmouth. Her specialties are: women's gender studies, the history of biblical scholarship, the history of Jewish-Christian relations, and anti-Semitism. One of her earlier books was *On Being a Jewish Feminist*. This book was a watershed for many Jewish women, but it also was a turning point for Heschel. She writes:

"With the publication of my book, I began to experience an emotional change as a feminist: my rage turned to laughter at the absurdity of sexism and its defenders, and I felt energized by the constructive efforts of those working for change. The extraordinary possibilities that were created for Jewish women—though hardly sufficient—meant that some of the horrors I had experienced would never confront future generations: there will never again be a time when a Jewish woman will be unable to find a minyan *(prayer quorum) that will welcome her to say* kaddish *(mourner's prayer)."*[79]

>> *Learn more about anti-Semitism:*
   http://www.ushmm.org/wlc/en/article.php?ModuleId=10005175

Blu Greenberg – Blu Greenberg (b. 1936) is an Orthodox Jew who has been active in the feminist movement for many years. Her father was a rabbi, and she was brought up in a traditional Orthodox home, where her father encouraged her to study Torah and often taught her himself. Her husband is also a rabbi devoted to the Jewish education of leaders. Greenberg received her M.S.

in Jewish history from the Yeshiva University in New York City. She has organized several international conferences on Judaism and feminism and was the first president of the **Jewish Orthodox Feminist Alliance**. She has also organized dialogues between Jewish and Palestinian women.

Early on, Greenberg began to question the exclusion of Jewish women in Orthodoxy. This questioning became more intense when she pursued the study of the Hebrew Bible in Israel and became more enlightened on these matters. Greenberg wanted to continue this study, but her parents objected, and she returned home.

Nevertheless, she became more and more dismayed by the fact that women could do nothing in the synagogue and was determined to be an advocate for change. Her passion stems from the belief that women have the same potential as men "whether in the realm of spirit, word, or deed."[80]

Ruth Bader Ginsburg - Ruth Bader Ginsberg (b. 1933) is a justice on the United States Supreme Court. She did her undergraduate work at Columbia University and studied law at both Harvard University and Columbia. At that time there was resistance to women being lawyers and Justice Ginsburg encountered discrimination. She decided then that she would be an activist for women's rights. In 1974 she wrote *Text, Cases and Materials on Sex-Based Discrimination*. She worked as a volunteer lawyer for the American Civil Liberties Union, specializing in women's rights, and argued many cases in the U.S. Supreme Court. She co-founded the *Women's Rights Law Reporter* in 1970, the first U.S. journal to deal only with women's rights. Justice Ginsburg was nominated for the Supreme Court by President Clinton—the second woman and the first Jewish female to hold that office. She has been outspoken in her support for abortion rights and for the equality of women.

SUMMARY

Judaism began when Abram answered God's call to form a new people. These chosen people eventually were led out of slavery in Egypt into the land of Canaan that had been promised them. There they began a history that included national glory, numerous conquests, dispersion, persistent recovery and reform, and, in our own time, near annihilation. Today Judaism stands as a significant world religion with its own State and a renewed commitment to witness the power of fidelity, devotion, and loving kindness.

The Hebrew tradition has been recorded and interpreted in the Torah and other writings in the Hebrew Bible, as well as in the Talmud and many other rabbinic writings. Interpretations of the tradition have differed, and Judaism divided into Orthodox, Reform, Conservative, and Reconstructionist. Even so, Jews have remained one people, one family. In the midst of many world cultures, Jews strive to be faithful to their belief in one God, their covenant with this God, commitment to Torah, and the celebration of Sabbath, High Holy Days, Passover, and other feasts.

Many Jews strive to sustain their threatened environment in Israel and be a force in the global ecological movement, bring a peaceful settlement to the conflicts in the Middle East and in other parts to the world, and honor the women's voices among them who call for equality. They know that their tradition enables them to bring rich treasure to the table for developing a global ethic and international cooperation for peace and justice.

One Jewish scholar says: "Today the Jewish population worldwide is larger than at any time in history. It is more integrated than ever before in its history in Western societies, in positions of leadership in such secular fields as education, business, medicine, and science. Jewish scientists have won Nobel Prizes in numbers far out of proportion to the size of the Jewish population. On the one hand, the Jewish diaspora has shrunk, and communities that dated back thousands of years have disappeared; on the other, the state of Israel has been reestablished after 2,000 years, and the Hebrew language reborn. Today the Jewish people are at a zenith in their history."[81]

CHAPTER 4 VOCABULARY

Covenant - A contract between Yahweh and the Hebrew people

diaspora - The dispersion of the Jews beyond Israel

Holocaust - The killing of the Jewish people by the Nazis during World War II

kosher - Food that is ritually correct

Midrash - Commentary on the scriptures by the rabbis

prophet - Someone inspired by God to speak to the people

rabbi - A religious teacher

Sabbath - The seventh day of the week: a day for rest and prayer

Seder - A ritual meal remembering the Passover

talit - Prayer shawl worn by men

Talmud - A vast commentary on the Hebrew scriptures

Torah - The first five books of the Hebrew scriptures

Yom Kippur - The Day of Atonement

Zionism - A movement to establish Israel as a nation

TEST YOUR LEARNING

1. Compare and contrast the significance of Abraham and Moses in the Jewish tradition.

2. How does the Torah differ from the Mishnah and Talmud?

3. What were the important changes in Judaism during the rabbinic period?

4. Why is Passover such an important celebration for Jews?

5. What are some of the important differences among Orthodox, Conservative, and Reform Jews?

SUGGESTED READINGS

Cohen, Shaye J.D. From the Maccabees to the Mishnah. *Louisville: Westminster John Knox Press, 2006.*

Cohn-Sherbok, Dan. Modern Judaism. *New York: Macmillan, 1996.*

Drucker, Malka. Women and Judaism. *Westport, CT: Praeger, 2009.*

Elior, Rachel. The Mystical Origins of Hasidism. *Oxford: Littman Library of Jewish Civilization, 2006.*

Heschel, Abraham. Maimonides. *New York: Farrar Straus Giroux, 1982.*

Kaplan, Dana Evan. American Reform Judaism. *London: Rutgers University Press, 2003.*

Rose, Or N. and others, eds. Righteous Indignation. *Woodstock, VT: Jewish Lights, 2008.*

Satlow, Michael. Creating Judaism. *New York: Columbia University Press, 2006.*

Schecter, S. Aspects of Rabbinic Theology. *Woodstock, VT: Jewish Lights, 1993.*

Sirat, Colette. A History of Jewish Philosophy in the Middle Ages. *New York: Cambridge University Press, 1985.*

Tirosh-Samuelson, Hava, ed. Judaism and Ecology. *Cambridge, MA: Cambridge University Press, 2002.*

NOTES

[1] *Elie Wiesel, The Trial of God (New York: Schocken Books, 1995).*

[2] *Jacob Neusner, Judaism: The Basics (New York: Routledge, 2006), 2-3.*

[3] *Darrell Jodock, Covenantal Conversations (Minneapolis, MN: Fortress Press, 2008), 19.*

[4] *Victor H. Matthews, A Brief History of Ancient Israel (London: Westminster John Knox Press, 2002), 12-13.*

[5] *Ibid., 15.*

[6] *Karen Armstrong, A History of God (New York: Ballantine Books, 1994), 7.*

[7] *Naomi Pasachoff and Robert J. Littman,* A Concise History of the Jewish People *(New York: Rowman and Littlefield, 1995).*

[8] *See Jean-Louis Ska,* The Exegesis of the Pentateuch *(Tubingen: Mohr Siebeck, 2009), 33ff.*

[9] *For a thorough exegesis of Exodus, see Thomas B. Dozeman, ed.,* Methods for Exodus *(New York: Cambridge University Press, 2010).*

[10] *Pasachoff, 15.*

[11] *Ibid., 114.*

[12] *John Bowker,* Jesus and the Pharisees *(New York: Cambridge University Press, 1973), 89-93.*

[13] *Ibid., 92.*

[14] *See Anthony J. Saldarini,* Pharisees, Scribes and Sadducees in Palestinian Society *(Edinburgh: T&T Clark, 1989).*

[15] *See http://dss.collections.imj.org.il.*

[16] *Neusner, 111.*

[17] *Ibid., 119.*

[18] *See Jacob Neusner,* Questions and Answers: Intellectual Foundations of Judaism *(Peabody, MA: Hendrickson, 2005), 6-49.*

[19] *Hans Küng,* Judaism: Between Yesterday and Tomorrow *(New York: Crossroad, 1992), 12.*

[20] *Pasachoff, 120.*

[21] *Barry W. Holtz,* Back to the Sources: Reading the Classic Jewish Texts *(New York: Summit Books, 1984), 136.*

[22] *Dan Cohn-Sherbok,* Judaism *(New York: Routledge, 2003), 138.*

[23] *John Corrigan, Frederick Mathewson Denny, et al.,* Jews, Christians, Muslims: A Comparative Introduction to Monotheistic Religions *(Upper Saddle River, NJ: Prentice Hall, 1997), 344.*

[24] *T. Carmi, ed.,* The Penguin Book of Hebrew Verse *(New York: Penguin, 1981), 348-349.*

²⁵ *See Daniel H. Frank and Oliver Leaman, eds.,* History of Jewish Philosophy *(New York: Routledge, 2003), 245-281.*

²⁶ *Louis Jacobs,* Principles of the Jewish Faith *(Northvale, NJ: Jason Aaronson Inc., 1988), 14.*

²⁷ *Cohn-Sherbok,* Judaism, *196.*

²⁸ *Ron H. Feldman,* Fundamentals of Jewish Mysticism and Kabbalah *(Freedom, CA: The Crossing Press, 1999), 32ff.*

²⁹ *Pasachoff, 137.*

³⁰ *Dan Cohn-Sherbok,* The Crucified Jew *(New York: HarperCollins, 1992), 72ff.*

³¹ *Paul Mendes-Flohr and Jehuda Reinharz, eds.,* The Jew in the Modern World *(New York: Oxford University Press, 1995), 97-98.*

³² *Lavinia and Dan Cohn-Sherbok,* A Short Reader of Judaism *(Oxford: Oneworld, 1997), 130.*

³³ *Cohn-Sherbok,* Judaism, *242.*

³⁴ *Nicholas De Lange,* Penguin Dictionary of Judaism *(New York: Penguin Books, 2008), 464.*

³⁵ *See Peter Neville,* The Holocaust *(Cambridge: Cambridge University Press, 1999).*

³⁶ *Elliot N. Dorff and Louis E. Newman, eds.,* Contemporary Jewish Theology *(New York: Oxford University Press, 1999), 371-372.*

³⁷ *See Oliver Leaman,* Evil and Suffering in Jewish Philosophy *(New York: Cambridge University Press, 1995), 185-220.*

³⁸ *Neusner,* Judaism: The Basics, *172.*

³⁹ *See Walter Laqueur,* A History of Zionism *(New York: Schoken Books, 2003).*

⁴⁰ *Holtz, 12.*

⁴¹ *See Neil Gillman,* Doing Jewish Theology *(Woodstock, VT: Jewish Lights Pub., 2008), 89-105. Roland Murphy,* Responses to 101 Questions on the Biblical Torah *(New York: Paulist Press, 1996).*

⁴² *Segal, 14.*

[43] See Roland E. Murphy, Ecclesiastes *(Dallas: Word Books, 1992).*

[44] Holtz, 183-184.

[45] Midrash Tanhuma, Ch. 9.

[46] Segal, 142ff.

[47] Richard L. Rubenstein, After Auschwitz *(Indianapolis: Bobbs-Merrill, 1966),* 42.

[48] David Hillel Gelernter, Judaism *(New Haven: Yale University Press, 2009), 167.*

[49] Gary A. Herion, ed., Ancient Israel's Faith and History *(Louisville: Westminster John Knox Press, 2001), 57ff.*

[50] Cohn-Sherbok, Judaism, 428.

[51] Neusner, The Way of Torah *(Encino, CA, Dickenson Publishing Co., 1974), 134.*

[52] Ibid., 140.

[53] See Robert Schoen, What I Wish My Christian Friends Knew About Judaism *(Chicago: Loyola Press, 2004), 69-116.*

[54] Hasia R. Diner, A New Promised Land: A History of Jews in America *(New York: Oxford University Press, 2003), 117ff.*

[55] See the website of the Anti-Defamation League, http://www.adl.org/about. asp?s=topmenu.

[56] Marc Lee Raphael, Judaism in America *(New York: Columbia University Press, 2003), 122.*

[57] Jeremy Benstein, The Way into Judaism and the Environment *(Woodstock, VT: Jewish Lights Pub., 2006), 6.*

[58] Ibid., 22.

[59] Ibid., 166.

[60] Ellen Bernstein, Ecology and the Jewish Spirit *(Woodstock, VT: Jewish Lights Pub., 1998), 14.*

[61] Vayikra Rabbah, 4,6.

[62] Lawrence Schiffman and Joel B. Wolowelsky, War and Peace in the Jewish Tra-

dition *(New York: Yeshiva University Press, 2007), 517.*

[63] *Peggy Morgan and Clive Lawson, eds.,* Ethical Issues in Six Religions *(Edinburgh: Edinburgh University Press, 2007), 203.*

[64] *Cohn-Sherbok,* Judaism, *566.*

[65] *See www.judaism.com/holiday/rosh_hashanah_yom_kippur/machzor.asp,* Machzor for Rosh Hashanah and Yom Kippur.

[66] *See http://www.emishnah.com,* Mishnah, Sayings of the Fathers, *1:2-18 in ml, 490.*

[67] *See American Friends of Neve Shalom/Wahat al-Salam: http://oasisofpeace.org.*

[68] *Neusner,* The *Way of Torah, 77.*

[69] *Jeremy Rosen,* Understanding Judaism *(Edinburgh: Dunedin Academic Press, 2003), 78.*

[70] *Arvind Sharma, ed.,* Women in World Religions *(Albany: State University of New York Press, 1987), 184.*

[71] *Judith Plaskow,* Standing Again At Sinai *(San Francisco: Harper and Row, 1990), 106.*

[72] *Rabbi Malka Drucker, ed.,* Women and Judaism *(Westport, CT: Praeger, 2009), 39-41.*

[73] *Sharma, 206.*

[74] *Dorff, 132.*

[75] *Drucker, 190.*

[76] *Dorff, 147.*

[77] *Drucker, 75 ff.*

[78] *Ibid.*

[79] *Jewish Women's Archive: http://jwa.org/feminism/_html/JWA036.htm*

[80] Jewish Women: A Comprehensive Encyclopedia: *http://jwa.org/encyclopedia/article/greenberg-blu*

[81] *Pasachoff, 336.*

Christianity

Magnus MacFarlane-Barrow, a salmon fisherman, was enjoying a pint in a pub with his brother in the Scottish Highlands in 1992. A news flash alerted them to people suffering in war-torn Bosnia. Magnus and his brother had been in Bosnia, had great memories of the hospitality, and wanted to help. They loaded their old Land Rover with food, clothing, medicine, and blankets and drove nearly 1,500 miles to Bosnia. When Magnus returned home he found so many donations that he had to continue helping others. After much prayer and thought, Magnus quit his job, sold his house, traveled back to Bosnia twenty-two more times, and then started a project called Mary's Meals, named after Jesus' mother. Magnus was inspired by Mary, whom he points out was a poor, single mother who had experienced exile and cruelty. Mary's Meals distributes 400,000 free meals a day to children in more than five hundred schools in fifteen countries. Magnus lives with his wife and six children and runs the operation out of a shed on his parents' property. He is a Christian of deep faith. His work harkens back to the words of a Jewish preacher who had taught other fishermen that they would possess the kingdom if they responded to those who say: "I was hungry and you gave me to eat" (Matthew 25:35).

Magnus' Christian religion began more than two thousand years ago with a small band of Jews who believed that Jesus, a Galilean preacher and healer, was the Messiah and their savior. This conviction galvanized them into spreading his teachings throughout Palestine, Syria, and then to the non-Jewish populations of Asia Minor, Greece, and as far as Rome. Eventually, the Christian religion attracted converts throughout the world. Today, it is the largest of the world's religions, numbering more than 2 billion people—more than one-third of the world's population. Christianity has quadrupled in the last century and is showing enormous growth today in Africa and Latino nations.

How did Christianity become the world's largest religion with traditions and beliefs that permeate many facets of world culture? To understand its emergence, growth, and establishment we will investigate Christianity's beginnings and the early centuries of its growth, and then we will explore the eras of its development, examining some of the religion's defining events. We will study the Christian Bible, especially the all-important four gospels, and then the Christian rituals and feasts. We will close by exploring global Christianity and some of the involvement of Christians with ecology, peace, and the women's movement.

ORIGINS AND GROWTH

The history of Christianity spans more than two thousand years. It is the story that begins with a person named Jesus of Nazareth calling a small group of disciples who grew their beliefs into a worldwide movement. It is the story of persecution, dominance, controversy, division, adaptation, and renewal.

Simple beginnings – Jesus of Nazareth is considered to be the founder of Christianity. We know little about Jesus' life, other than what has been written in the gospels, which were composed decades after his death. There have been many scholarly attempts to describe the "historical Jesus." These efforts offered little return in the nineteenth century, but in more recent studies, archaeological as well as historical, social studies and biblical research have enabled scholars to propose "portraits" of Jesus. These portraits, of course, vary considerably.

Many scholars think that he did not set out to establish a new religion. Instead, Jesus seems to have been a Jewish reformer, passionate about renewing the best Jewish values of love, kindness, prayer, and self-sacrifice in his own Jewish religion.[1]

According to the gospels, Jesus was raised in the small village of Nazareth by his parents, Mary and Joseph. Little is known of his childhood. Jesus was apparently trained by his father to be a craftsman, working with stone and wood. We have no pictures of Jesus, but judging from what we know of Jews of the times, he would have had dark eyes, hair, and skin. Most likely Jesus had little education. He spoke Aramaic and knew enough Hebrew to read the Torah in the synagogue, but for the most part was illiterate, having no access to written materials. At the same time, Jesus seems to have been gifted, because his later teachings, though simple and often given in **parables (stories that teach about faith)**, were seared into the memories of his listeners and are still quoted today.

The Ministry of Jesus – Most of what we know of Jesus comes from the **gospels**, the first four books of the New Testament. Gospels are literary creations, based on what Jesus said and did, but written many years after Jesus' time. Many biblical scholars think that these stories are more concerned with beliefs and faith than they are with history.

Mark's gospel, the earliest of the four gospels (ca. 70 CE), tells of Jesus' baptism by John the Baptist. During the baptism, Jesus recognized that he was beloved of God. Afterward, Jesus spent forty days in the desert fasting and facing serious temptations. Then, Jesus returned to Galilee and began his ministry with his central message: "The time is fulfilled, and the kingdom of God is at hand. Repent and believe the gospel" (Mark 1:15). He calls people to change the direction in which they are seeking happiness, to turn their lives toward goodness and love, to practice the spirit of Jewish law and not just the mechanics. He invites them to participate in bringing about the reign of God's love and mercy in the world.

The gospels remember Jesus as "meek and humble of heart," as one drawn in compassion to the poor and disabled, one who was often moved to bring forgiveness and miraculous healing to people. The gospels tell many stories about Jesus working miracles (acts of God's divine power), as he heals the blind, the crippled, the diseased, and even raises the dead to life. Though meek and humble, Jesus is also remembered as being confrontational. He had little tolerance for the self-righteous or hypocritical—those who thought they could be saved by simply obeying the Jewish laws, yet not act in accordance with the spirit of their faith. The gospels describe Jesus' controversial engagement with some of the **Pharisees**, Jewish leaders who saw themselves as role models for religious behavior. He also encountered **scribes**, who were the Jewish religious scholars of the time.

The gospels depict Jesus challenging Herod, who was appointed by the Romans as a Jewish king and collaborator of the area, considered a Roman client state. The Romans had conquered and subjected Jewish people about fifty years before Jesus was born. He confronted the greed and corruption of the Temple moneychangers, tax collectors, and merchandisers who fed the coffers of Rome. He confronted their appointed puppet leaders among the **Sadducees**, wealthy and powerful Jewish nobles. Jesus was possibly disgusted by their willingness to support Roman domination as long as they could live "the good life" comfortably in their villas near the Temple. Such actions created enemies.

It was inevitable that Jesus would be perceived to be dangerous by many of the Jews and Romans in authority. He had unveiled their corruption and threatened their power. Jesus had given hope and courage to the poor and outcasts, and his miracles had drawn crowds who were amazed at his gifts.

Jesus had chosen a small band of men and women and trained them to carry on with his preaching and healing. Jesus chose twelve apostles, who would be leaders in his new movement. He also chose many disciples who would be commissioned to continue his ministry of teaching and healing. Jesus promised that he would always be with them "in spirit" as they carried on his work. He was unique among Jewish teachers in selecting women disciples, chief among whom was Mary Magdalene. She was a woman of means who helped support the movement. There is no evidence that she was ever a prostitute. Jesus chose most of his followers from the common people, fishermen and laborers, and from those held in disdain by religious Jews, such as tax collectors and the unclean. The gospels tell of the emergence of Peter as a leader with Jesus' words: "Now I say to you that you are Peter (which means 'rock'), and upon this rock I will build my church, and all the powers of hell will not conquer it" (Matthew 16:18).

Some of Jesus' central teachings seem to be summarized in Matthew's Sermon on the Mount (Matthew 5—7) and in Luke's Sermon on the Plain (Luke 6:20–40). Jesus exhorted his followers to love their enemies, avoid judging others, and avoid anger, adultery, divorce, false oaths, and revenge. He urged them to give alms to the needy and to be dedicated to prayer and fasting. Jesus taught them to store treasures in heaven rather than on earth, to depend on God, and to follow the Golden Rule: "Do to others whatever you would have them do to you" (Matthew 7:12).

Jesus' central teaching is one of love. When the Pharisees asked Jesus what the greatest commandment was, he answered: "You shall love the Lord, your God, with all your heart, with all your soul, and with all your mind." Then Jesus named the second command as "You shall love your neighbor as yourself" (Matthew 22:37–40). Jesus drew both of these laws from the teachings of his Jewish Scriptures.

After a final Passover supper with his disciples, Jesus gave them a simple, ritual meal and asked them to celebrate this meal in his memory in the future. Jesus was betrayed while praying in a garden by one of his own, Judas Iscariot. The gospels tell how he was arrested, tried on false charges, and condemned to death by crucifixion.

With great dignity, Jesus accepted beatings, scourging with thongs of leather and sharp iron, and torture with a crown of thorns. He refused to back down on his mission and message and remained defiant toward both the Jewish and Roman leaders. In the gospels, Jesus even dared to face down the powerful Roman Procurator, Pontius Pilate, who alone held the authority to crucify him. Jesus was led off carrying the heavy log that would be the horizontal piece to his cross. He was nailed to the cross and left to die naked between two criminals.

Apparently it was over. The short, amazing mission of this young Jew from Galilee had come to a humiliating and tragic end. Most of his followers, aside from a few devoted women, fled and hid lest they too be executed. His betrayer, Judas, hanged himself. The enemies of Jesus, both Jews and Romans, could feel relieved. They had silenced this crude, young upstart who exposed their greed and hypocrisy. This peasant carpenter had the nerve to challenge their harsh and often violent misuse of power with his life and teachings filled with humility, love, and compassion.[2]

The gospels recount how Jesus' disciples were startled by a confounding event that happened early on the morning of the third day after his crucifixion. Some women came to anoint the dead body, found his tomb empty, and were told that he had been raised from the dead. The gospels proceed to give accounts of various appearances the risen Jesus made to his disciples after his apparent **resurrection**. These appearances create a turning point in the lives of Jesus' followers. They now begin to recognize Jesus to be the Christ, the anointed one of God, the Messiah that the Jews had hoped for, the Son of God whom they could now worship. The gospel of Mark closes with the account of Jesus' **ascension** or return to the heavens. Jesus then commissioned his disciples to teach his gospel to the whole world. Their hope and courage to continue on with this "dangerous message" came from his earlier promise that he will be with them "all days even to the end."

The Message Spreads – *The Acts of the Apostles*, which was probably written in the 80s CE, gives an account of the birth of the Christian Church on **Pentecost**. The Greek word "Pentecost" means "fifty days." It was now fifty days since the resurrection of Jesus. Traditionally, this has been considered to be the birthday of the church. The community had gathered in Jerusalem, when they were suddenly filled with the Holy Spirit (the divine Spirit of God). Peter, the lead apostle, stands before a large crowd and proclaims: "Men of Israel, hear these words. Jesus of Nazareth was a man approved by God among you by miracles and wonders and signs, which God did through him in the midst of you." Peter goes on to say that Jesus was crucified, but that "God raised him up…" (Acts 2:22–24).

The early followers of Jesus continued to spread his message and perform healings. They followed what they called "the Way," and lived in communities where they called each other brother and sister. They shared what they had in common. They made an attempt to remain in the synagogues and the Temple, but these followers were soon turned away because of their insistent belief in Jesus as the Messiah.

Small communities began to appear in Antioch and Damascus, in Syria and in Corinth, Greece. Communities even appeared in the capital of the Roman Empire. Within ten years after the death of Jesus, a Pharisee named Paul, according to the Scriptures, became converted by an experience of the risen Jesus. He dedicated himself to bringing the Christian faith to the **Gentiles (non-Jews)**. Standing up to Peter, the prime apostle, and James, the brother of Jesus and head of the church in Jerusalem, Paul persuaded them that Gentiles who converted would not have to be circumcised or live as Jews. Paul spent the next thirty years spreading the faith among non-Jews, traveling toward the east in Asia Minor (now present-day Turkey) and Greece.

In his letters to some of the communities, Paul developed a brilliant theology about Jesus' divinity, the meaning of the cross, and even Christ's cosmic significance. This theology would leave a powerful stamp on Christianity and later would be included in the official Scriptures. Paul sums up the early Christian belief in Jesus' divinity and saving power through the cross when he writes to the Galatians: "I live by faith in the Son of God, who loved me and gave himself up for me" (Galatians 2:20).

Persecution – Fledgling communities, sometimes under attack from hostile Jews or Romans, met secretly in house churches. The Scriptures recount how Stephen was the first to be martyred in Jerusalem. Later, many were killed by the Emperor Nero in Rome, including Peter and

Paul in the mid-60s CE. Under later emperors, such as Decius (249–251) and Diocletian (284–305), Christians were executed in great numbers, accused as being atheists for not adoring the emperor or traitors for refusing to participate in violence as Roman soldiers. A devotion toward these martyrs and their relics spread throughout the Christian community. Tertullian (ca. 155/160–220), an African theologian, wrote the classic statement, "the blood of the martyrs is the seed [of the church]."

Early on, the secrecy of the Christians led to vicious rumors. It was said that Christians consumed the body and blood of Jesus in **Eucharist** (receiving the bread and wine). This led to the rumor that they were cannibals. Their reference to each other as brother and sister led to suspicion of incest. And when earthquakes or destructive storms came, some Romans blamed the Christians for displeasing the gods with their false beliefs, such as holding that Jesus was their divine Lord rather than the emperor. Pliny the Younger, a Roman official, spied on Christians and gave his report at the beginning of the second century to Emperor Trajan. Pliny wrote that the Christians were practicing a "perverse superstition," and he observed that "they were in the habit of meeting on a certain fixed day before it was light, when they sang in alternate verses a hymn to Christ, as to a god, and bound themselves together by a solemn oath, not to do any wicked deeds.... after which it was their custom to separate, and then reassemble to partake of food." [3] Most likely this secret meal was "the breaking of the bread," which was the sacred meal of the early Christians, handed down to them by Jesus. **Baptism** was their ritual to initiate candidates into the community. Usually it involved submerging the candidates, symbolizing the death of Jesus, and then raising them from the water to symbolize entering into the resurrection of Jesus.

THE EARLY CENTURIES

The Jesus movement continued to spread, partly due to the courage and fervor of the converted Gentiles who were now outnumbering the Jews in

Christianity. Many who were fleeing from persecution spread the gospels as they traveled. By the end of the second century, Christian communities could be found in France, Germany, Spain, Britain, North Africa, Persia (now present-day Iraq and Iran), and even India. It is estimated that by 300 CE the Christians numbered more than 5 percent of the population of the Roman Empire, and the religion was expanding rapidly.[4] A clerical system of bishops, priests, and deacons gradually evolved, and an authority structure was established.

Christianity became a largely urban phenomenon. The bishops of the chief cities of Antioch in Syria, Alexandria in Egypt, and Rome held the key positions. Rome was the location of the capital of the empire and the burial place of both Peter and Paul; therefore, the bishop of Rome began to gain more influence. It was out of this office that the papacy ultimately developed.

In the early fourth century, the Christians had just experienced a serious persecution. The Roman Empire was divided with competing emperors struggling for power. There was a showdown at the Milvian Bridge just north of Rome, and the Roman emperor Constantine (ca. 272–337 CE) was one of the contestants. He claimed to have experienced a vision of the cross before the battle and as a result was victorious. Constantine converted to Christianity and then published the Edict of Milan in 313, which declared that Christianity was now free to operate within the empire.

This was an amazing turnaround for the Christians. No longer would they be outsiders and have to suffer persecution at the whim of the emperor. Imperial funds now flowed into the church, bishops and priests were tax exempt, and church services could be publicly celebrated. Large Roman basilicas (halls of justice) were donated for church use, and Christians began to build their own churches modeled on these plans. Elaborate services were developed with the celebrants dressed as Roman nobles.

Eventually, Constantine also conquered the Eastern Empire and in 324 set up Constantinople (now present-day Istanbul) as the capital of the empire and Christianity. Constantine's mother journeyed to Palestine and claimed to discover the sites where Jesus was born and died, as well as the true cross. Constantine erected basilicas on the holy sites and transformed Jerusalem into a Christian city, a place for solemn pilgrimage. It would remain so until it was briefly taken over by the Persian Zoroastrians, a religious group, in 614 and then conquered by the Muslims in 638.

The next few decades saw some struggle for power between the non-Christians and Christians, but in 380 CE the Emperor Theodocius declared Christianity to be the official religion of the empire. Now the non-Christians would be the ones to be persecuted.

Throughout the third, fourth, and fifth centuries the western regions of the Roman Empire were attacked by the barbaric tribes from the north—Huns, Goths, Vandals, Burgundians, and Lombards. Rome was sacked several times in 410 and left desolate by King Alaric and the Visigoths. The Western Empire's center had to be shifted to Ravenna, Italy, inland from the Adriatic Sea. The last Roman emperor of the Western Empire, Romulus Augustulus, was deposed in 476, marking the end of an empire that had spanned nearly one thousand years.[5]

Amidst all this chaos, the church in the West struggled to survive amid violence and famine. At the same time, in the power vacuum, popes and bishops gained influence and were the leaders people turned to for food, clothing, and health care. Their service was impressive, and their example and leadership helped move many in the barbaric tribes to embrace the Christian faith.

While the church in the West floundered and entered the so-called Dark Ages (476–800), the Eastern Church flourished. The Emperor Justinian (ca. 483–565) strengthened the East and built the magnificent Hagia Sophia, also called the Church of the Holy Wisdom, in Constantinople, which still today stands as one of the world's greatest religious structures. Justinian also decided to take over the Western Empire and successfully laid waste to the barbarian cities across Europe as far as the interior of Italy. But his plans were thwarted by the appearance of the bubonic plague that leveled half of Europe's population during the mid-sixth century.

Early Councils – The Roman Emperor Constantine had sought to use the widespread and well-organized system of Christianity as the glue to hold together his fragile empire. He was frustrated when he discovered that his unifying factor was itself torn by controversy. The main tenets of Christianity had been broadly accepted:

- the Incarnation, or the belief that God became human in the form of Jesus
- Jesus' humanity and divinity
- the very nature of the Holy Trinity (three Persons in one God: Father, Son, and Holy Spirit)
- the power of the cross to save the world
- the belief in resurrection to eternal life after death

But many of these beliefs would now be challenged and hotly debated.

From the beginning, there had been conflicting beliefs in the church, but the most persistent heresy that the church or Constantine had ever dealt with was called Arianism. The originator of the controversy—Arius, a brilliant priest—taught that the transcendent God could not have possibly entered into the created world as a man. Therefore, Arius concluded that Jesus must have been created by God as a higher being and could not be divine.[6]

In 325 CE, Constantine called to order the Council of Nicea with three hundred bishops. He presided over their gathering and gave them orders to settle the Arian controversy. The Council was clear in its final decree, one that is still recited in some churches today:

> *"We believe in one God the Father Almighty, Maker of all things visible and invisible; and in one Lord Jesus Christ, the only Son of God, eternally begotten of the Father, God from God, Light from Light, True God from True God, Begotten, not made, of one substance with the Father, through whom all things are made."*

Arianism wasn't defeated, but as far as most of the bishops were concerned the controversy over Jesus' divinity was settled. Anyone denying it was declared to be a heretic or one who rejected a doctrine of the church.

The debate over Jesus' humanity and divinity continued between the two great churches in Antioch and Alexandria. The fiery Cyril of Alexandria had Nestorius the Patriarch of Constantinople (386–451) condemned and exiled because Nestorius taught that Mary gave birth to the human Jesus, but not to the divine Son of God; he split the two natures of Christ. Finally, the Council of Chalcedon in 451, amid shouting matches and even the occasional fistfights, settled the matter about the two natures of Jesus. The Council defined that Jesus Christ was "truly God and truly man," with two natures united in one person in complete unity.

Augustine Confronts Heresies – There were other doctrinal disputes that divided Christianity at the turn of the fourth century. The Donatists held that unworthy ministers were not able to celebrate valid sacraments. The Pelagians denied that the fall in the Garden of Eden corrupted human nature, and thus taught that people could be saved without the grace of God.

Augustine (354–430), one of the greatest thinkers in Christian history, was called upon to confront these challenging notions. Augustine had been a scholar and teacher for years. For more than ten years, he lived with his common-law wife and their son. The violence of the Old Testament repelled him, and he joined the Manicheans, a Gnostic community, which took a dim view of human nature and encouraged intense asceticism. Augustine was eventually converted to live the Christian life. He left his partner and child, formed a monastic community, was ordained a priest, and was chosen to be bishop of Hippo in North Africa. He followed his now classical observation: "Our hearts are restless until they rest in Thee."

>> *As you read these quotations from St. Augustine connect them to tenets of the Christian faith as taught by Jesus:*
http://www.youtube.com/watch?v=rMxnUsWgXqw&feature=related

Augustine spent the rest of his life ministering, preaching, and writing. He produced a number of Christian classics. Among them are his autobiography *Confessions*, as well as *The City of God* and *The Trinity*. Influenced by positions in the Roman Empire, he also developed the *Just War Theory*, still in current use, which lists the conditions necessary to justify going to war.

Augustine attacked the Donatists and showed that since sacraments (the sacred rituals of the church, e.g., baptism and Eucharist) are celebrated by the power of Christ, ministers who preside can be unworthy. He developed a theology of sacraments that is still highly influential today. Against the Pelagians, Augustine developed the well-known Christian theology of *original sin*, demonstrating how humanity was corrupted by the fall from grace in the Garden of Eden (Genesis 3:1-24). Augustine insisted that people confess their sinfulness and aspire to become pure like Jesus. He was adamant that God's grace was needed for salvation. Augustine also developed the mainstream theology of the Trinity, teaching that the One God was manifested as Creator (Father), Son (Jesus Christ), and empowering Spirit (Holy Spirit). Augustine maintained that the Spirit was breathed forth from the Father and the Son. As we will see later, his theology of original sin and grace would influence Martin Luther, and Augustine's view on predestination was taken up by John Calvin, a Protestant reformer.[7]

Monasticism – The Christian movement in the third and fourth centuries, for the most part, existed in cities and was still surrounded by non-Christian customs, political strife, materialism, and the usual corruption associated with urban areas. Many Christians, wanting to live Christianity more fervently, left the cities for the deserts, where they could live as hermits in silence and solitude, depriving themselves of the things of "the world." This lay movement in the Eastern Church began in the deserts of Syria and Egypt. Its most famous leader was Anthony of the Desert (ca. 251–356). Early in the fourth century, Anthony—a converted soldier—reacted against some of the extreme forms of fasting and self-abnegation of the hermits. He drew up more moderate rules for communities, first for men and then for women, where monastic life could be lived devoutly. The movement was improved further by Basil the Great (330–379), who established a flourishing community in the extraordinary rock formations of Cappadocia, Turkey. The amazing caves where the monks lived and provided education, health care, and relief for the poor can still be visited by pilgrims today.

In the West, monasticism developed under the leadership of Cassian (360–435) in France and Cassiodorus (490–583) in Italy. It was Benedict (480–547) who wrote his famous *Rule for Monks* that gives every detail for living a life of prayer and work in a monastery or convent and who established the still existing monastery at Monte Cassino in Italy. This monastery set the standards for monastic life in the West, a life governed by prayer, hard work, and the vows of poverty, chastity, and obedience.

Monasticism also flourished in Ireland, where a distinctive Celtic approach to Christian spirituality was developed. Celtic spirituality honored nature and settled in "thin places," lovely green and water areas where God could be experienced intensely. Monasteries provided refuge from the barbarian tribes in Europe. The Celts also showed great respect for the feminine and offered women active participation in the rituals. It has been said that these monks preserved the civilization of the West by copying and preserving the classic manuscripts of antiquity, writings that would later be foundational to the establishment of medieval and enlightenment culture and civilization.[8] One of the most famous of these Celtic monasteries was founded by Columba (521–597) on Iona, an island off the coast of Scotland. This monastery brought about many conversions between the Scot and Pict tribes. Today, visitors may brave the waves of the Irish Sea and visit the remains of the monastic, beehive-shaped huts on the beautiful Skellig Islands.

》 *Listen to Gregorian Chant Benedictinos; then read the English translation of this prayer to Mary the Mother of God. How does experiencing the prayer in song differ from reading it?*
http://www.youtube.com/watch?v=_MbDqc3x97k&feature=related

THE EASTERN AND WESTERN CHURCHES

The struggle for power between Eastern and Western Christianity was personified in the emperors of Constantinople and the popes in Rome. The East claimed to be heirs of Jerusalem and the true cross, while the West claimed to be the heirs of Peter the prime apostle, personified in the bishop of Rome, the pope.

Rome and the Western Church in the sixth and seventh centuries saw itself as more powerful than the East, which had been weakened by a prolonged and violent controversy over devotion to icons.[9] The Eastern Empire fought off a number of heresies and was also struggling to fend off attacks by the Persians and Muslims. By the seventh century Islam had conquered Syria, Palestine, Egypt, North Africa, and the greater part of Asia Minor, all of which from ancient times had been strongholds of Christianity. In 716 the Muslims even conquered parts of Spain. By 800 the great patriarchal Eastern Churches—Alexandria, Antioch, and Jerusalem—were under Muslim rule, and the patriarchs had to live in exile in Constantinople (now present-day Istanbul, Turkey). There was a serious breach between Constantinople and Rome in the mid-ninth century and a definitive split between the two churches, East and West, in 1054. On the morning of July 15 of that year, three papal legates stormed into Mass at Basilica Hagia Sophia, leveled accusations against the attending Patriarch of Constantinople, Michael Celarius, and placed a bull (a papal document) of excommunication for heresy on the altar. The Patriarch was understandably astounded and soon called a council of bishops that in turn excommunicated the pope and his legates. East and West were divided and remain so even today, though recent popes have attempted to bring about reunion.

The split between East and West evolved throughout the centuries. There was a political break in 800 when the West recognized Charlemagne as emperor of the Holy Roman Empire; this was perceived by the East as a break with the Byzantine Emperor. Theologically, the division arose from the West's reliance on Augustine, while the East turned to fathers of the Councils: Gregory, Basil, John Chrysostom, and Cyril of Alexandria. Much was also made about the Trinitarian controversy—whether the Holy Spirit proceeds from the Father and the Son (the West) or only from the Father (the East). There was also division regarding celibacy for priests and the type of bread used for celebration of the Eucharist. But the greatest cause for division was the Western Church's assertion of the universal authority of the pope.[10]

The Eastern Church – In 1254 the Western crusaders captured Constantinople, devastated the city, brutally dealt with its people, and tried to establish the Western Empire there. Constantinople regained control and held off Muslim forces until 1453, when it was overrun by the Turks. At that point, the center of the Orthodox Church shifted to Russia, where it became autonomous. The other Orthodox churches spread through the Slavic, eastern European countries and the United States, where they still exist today.

The Eastern Church flourished in the Slavic countries and in Russia until the Russian revolution in 1918 and the takeover of Eastern Europe after World War II. The Communists confiscated much of their property. The church's hierarchy, clergy, and nuns were persecuted, exiled, and killed. It has been estimated that from 1917–1943 the Communists killed more than three hundred bishops and 45,000 priests, along with hundreds of thousands of lay people who were tortured, sent off to concentration camps, or killed.[11] The fall of the Soviet Union saw the liberation of these churches and the return of their property. However, it will take many years to rebuild the churches, educate the clergy and people in the Christian faith, and restore order to the church organization.

The Eastern churches have prided themselves on being the descendants of the Greeks, who were converted by Paul and other apostles. They are also proud that they use Greek, the very language of the Christian Scriptures. They believe that they are carrying on the faith and rituals of the early church. Their majestic services are elaborate with magnificent music and singing, as well as processions around the altar with incense to honor the Lord Jesus. The splendid icons of the Eastern churches depict Jesus, Mary, and the saints. These icons are not simply pictures, but elaborate likenesses that are believed to make these revered ones present for veneration and petitions.

The Eastern churches still maintain ancient monasteries at much-honored places such as Mount Athos in Greece and the monastery of Saint Catherine at Mount Sinai, centers of great learning and intense spirituality. Eastern Christianity has its own unique approach to mysticism and has produced many outstanding saints, including John Chrysostom, bishop of Constantinople and a famous preacher (ca. 349–407), Cyprian, the bishop of Carthage (ca. 200–258), and Thekla (d. 100), who worked with Paul to spread the gospel and was considered an apostle.

Eastern theology differs considerably from the theology of the West in that it commonly does not use theoretical and logical structures. Rather, Eastern theology is spiritual, mystical, and poetic. Scriptures and the early writings of the Church Fathers provide main sources for reflection for both clergy and lay people.

Eastern theology is often concerned with a God who is beyond comprehension, the eternal and mysterious Trinity. This Trinity consists of: the Father, the eternal Source; the Son, who became the Christ and in this incarnation is both fully human and fully divine; and the Spirit, who finds its source in the Father. Eastern theology is centrally concerned with Jesus Christ, the Messiah and savior, the Risen Lord whose incarnation restored humanity to being made in the image and likeness of God. From this unique perspective, salvation is uniquely achieved by Jesus' incarnation, as well as by his life, death, and resurrection. This devotion to Jesus has been captured by the nineteenth-century Russian poet A.K. Tolstoy (1817–1875), whose poem was later set to music by composer Pyotr Ilyich Tchaikovsky:

> I behold him before me
> With the crowd of poor fishermen;
> With quiet, peaceful steps
> He walks between the ripened bread;

He pours the joy of his benevolent words
Into simple hearts;
He leads his flock, hungering for his righteousness,
to its waters.
Why was I not born at the time
When he was among us in the flesh,
walking life's path
and bearing his agonizing burden![12]

The ultimate goal for humans in Eastern spirituality is to become like God, as stated in their ancient saying: "God became human so that humans could become godly." To those who might ask how is it possible to become like God, Eastern scholars would answer that even though God is in essence unknowable, God's divine "energies" can be perceived and experienced. Eastern theology is characterized by the search for wisdom and is deeply concerned with living a life that is spiritual. Such a goal requires serious spiritual training and prayer, but is accessible to all Christians, not just monks and nuns.

The Eastern churches include the Uniate churches, which have remained in union with Rome and are made up of the Armenian, Chaldean, Maronite, Melkite, Syrian, Greek, and other churches. Although still loyal to Rome, the Uniate churches often complain of the lack of recognition from the dominant Roman Catholic Church.

The Orthodox churches, which include the large Greek and Russian churches, as well as other churches, number about 200 million Orthodox Christians worldwide. With the collapse of the Soviet Union there has been a strong revival of Orthodox Christianity in Russia, Siberia, Ukraine, and Kazakhstan. In these areas, the Orthodox Church has regained control of many of its churches and, as mentioned earlier, is struggling to rebuild their communities, facing much difficulty and lack of resources. These churches are reclaiming their commitment to doctrine, unique spiritual theology, elaborate ceremonies, and special traditions.

They still face serious divisions among themselves, especially among ethnic groups. The Orthodox churches are often adverse to the intrusion of evangelical groups from foreign countries.[13] The Orthodox communities in America experience division among the Coptic, Syriac, Greek, Ukrainian, and other groups. Each is strong in its own right, but faces the challenges for unity among themselves and adapting to the American culture. There are 2 million to 3 million Orthodox Christians in two thousand parishes in the United States.[14]

After nearly a century of persecution, many Orthodox leaders express a new optimism about the future. Alexei the Patriarch of Moscow and all of Russia comments: "Orthodoxy is one of the few religious confessions whose membership is growing rather than declining. After many decades of persecution, a major revival of spiritual life is underway in Russia and in other countries of the former Soviet Union, and it brings us joy that the number of parishes, monasteries and theological schools is significantly increasing."[15]

The Church in the West – In the West, **Christendom** was the name given to the church-state of Europe consisting of the ecclesiastical hierarchy and secular rulers. Each was to have authority over their respective areas, but in reality the two groups were often in conflict. The power of the Western Church was greatly enhanced by an extraordinary pope, Gregory the Great (540–604). A former monk, Gregory, in the absence of imperial power, managed the political, military, and religious life of Rome. He organized the Mass, collected chants, now known as Gregorian chant, and set up health care and other services for the poor during the plague. Gregory also sent Augustine of Canterbury to convert England.[16]

In 500 Clovis, the warrior ruler of the Franks, converted to Christianity, and a strong bond was created between the papacy and Frankish leadership. By the eighth century, Celtic missionaries, as well as Boniface, an English missionary and reformer, were making converts in Germany.

The church's Frankish alliance continued with Pepin, king of the Franks, and then with powerful Charlemagne, who was crowned as the Holy Roman Emperor. The West had stopped the Muslims in 732 in France, and the empire was secure. Charlemagne encouraged the copying of classical manuscripts in the monasteries and commissioned the talented Alcuin, an educator and cleric, to set up excellent cathedral schools throughout the empire. These became new centers for Christian education and scholarship.

Meanwhile, the faith continued to spread through the West. In the ninth century, Anskar brought Christianity to Denmark and Sweden, and Cyril and Methodius missioned to the Slavs. In the tenth century, Poland, Hungary, and Norway were converted. Churches in the expanding Christendom now replaced what had been lost in the Eastern churches.

The Crusades – In the early Middle Ages (1066–1307) there was much Christian animosity toward Islam. Muslims were viewed as heretics since they did not accept the final prophetic role or divinity of Jesus Christ. Though the Muslims had been tolerant of the Christians they conquered, the great centers of Christianity in the East no longer existed. Religious leaders in Rome were rankled by the Muslim control of the Holy Land and their construction of a mosque over the ruins of Herod's Temple. The Muslims were seen as enemies of the faith, and so in 1095 Pope Urban II called for a Crusade to liberate the Holy Land from Islam. The nobles were attracted to the idea because this would be a "holy war," offering opportunities for travel to exotic lands, for adventurous battles, and for securing bounty and land grants. Another motivation was spiritual purification and heavenly reward.

The First Crusade was a brutal victory. The Christian armies captured Jerusalem, slaughtered great numbers of resident Jews and Arabs, and set up a Christian kingdom whereby the pope controlled the holy places. The following Crusades were failures. By the end of the twelfth century, the Holy Land was back in the hands of the Muslims.

THE MIDDLE AGES

Some have said that the medieval period (1066–1485) was one of the greatest periods in Christendom. It was the time when the splendid Gothic cathedrals such as Durham and Canterbury in England, and Chartres and Notre Dame in France, were built, and the period when magnificent universities flourished in England, France, and Italy. Dante of Italy wrote one of the best poems of all time, *The Divine Comedy*, and there were brilliant teachers such as Abelard, John Duns Scotus, and Thomas Aquinas. Aquinas has been described as "one of the greatest masters of Christian thought, an exceptional moment in the chain of tradition."[17] Aquinas had a firm grasp of the great Christian thinkers and was also able to integrate the philosophy of Aristotle into his theological interpretations of the Christian tradition. His *Summa Theologica* is a brilliant classical synthesis and interpretation of the Christian tradition that has influenced numerous theologians until the present day.

This period also witnessed remarkable saints, such as Francis of Assisi, who brought Christians back to the simplicity and love of Jesus and is still honored worldwide. Francis was dedicated to a gospel life of poverty and service and founded a religious order that carries out his ideals today.

During the medieval period, the dramatic Cluny reform of the Benedictine monasteries spread throughout Europe. The monastery of Cluny in Burgundy had been founded in 910 and was directly under the patronage of the pope and, therefore, free of any local lay or episcopal control. The Cluny reform allowed more than one thousand monasteries to affiliate with Cluny, pay a small tax, obey the rules of Cluny, and thus be in complete control of their lands and revenues. This greatly strengthened the monasteries as well as the power of the papacy.

During this period, the papacy grew stronger, especially with popes such as Innocent III, an astute political and religious leader, who brought many reforms to the church. Innocent reformed the administrative structures in Rome, gained central control over the bishops, and achieved much influence over the monarchies of Europe. Blots on his rule were his isolation and mistreatment of Jews and his extreme position that the pope was semi-divine and Lord of the world.

>> *Observe the great Gothic cathedrals of Europe and connect the architecture to the Christian symbols:*
http://www.youtube.com/watch?v=05Z1gxGhRQQ

Christian Mysticism – From the very beginning of Christianity there has been a mystical tradition, a belief that disciples of Jesus can have a direct experience of God, which goes beyond everyday Christian life and prayer. We see an example of this in the gospels when the apostles Peter, James, and John experienced Jesus transfigured before them on a mountaintop (Mark 9:1–8). The Acts of the Apostles recounts when Paul the Apostle experienced the risen Jesus with such force that he was knocked down (Acts 9:1–9). Augustine speaks of his vision of God; Aquinas compares all of his writings as so much "straw" compared with his mystical experiences of God.

In medieval Italy, Francis of Assisi (1181–1226) experienced God in the poor, as well as in animals and nature. In England, the author of *The Cloud of Unknowing* taught that God comes to us when we go beyond reasoning, in the clouds of doubt and confusion as a light breaking into our consciousness. English mystic Julian of Norwich (1342–1416) wrote of her experiences of Jesus and how he taught her about God's love and compassion at a time when some were blaming God for the horrors of the Black Death. In Germany, Hildegard of Bingen (1098–1179), a brilliant Benedictine abbess, consultant to popes and kings, and composer of inspiring church music, wrote how her visions led her to more deeply appreciate the nobility of being human. (Her life was recently made into a movie called *Vision*.) Also in Germany, the priest Meister Eckhart (1260–1327) recorded his experiences of God in such a riveting manner that they are still read today. In Renaissance Spain Carmelites Teresa of Avila (1515–1582) and John of the Cross (1542–1591) produced mystical writings that are classics and still highly influential today.

The Inquisition – The Inquisition was a court system set up to judge and pass sentence on heretics. The Roman Inquisition was established in the thirteenth century by Pope Gregory IX (1227–1241), who appointed an Inquisitor, usually a Dominican friar, to seek out those who were in heresy. They were brought before the Inquisitor and if found guilty were sent to prison or burned at the stake. In the 1480s the Spanish Inquisition was established to pursue Jews and any Christians suspected of heresy. This Inquisition was later extended to Latin America and justified the flogging and even execution of heretics, Protestants, and indigenous people.

THE PROTESTANT REFORMATION

The Protestant Reformation of the sixteenth century caught many by surprise because in certain ways Christianity in the West seemed to be doing well at the time. The church in England was strong, and its flamboyant King Henry VIII had been named "Defender of the Faith." The papacy was formidable and could face down most nobles and monarchs with threat of **excommunication** (exclusion from all the rights of the church) or with **interdict** (the withholding of certain sacraments). The New World had just opened and held great potential for wealth and vigorous missions. Europe was experiencing a renaissance, complete with publishing from the new printing presses, and amazing paintings and sculptures produced in Florence, Milan, and Rome. The wondrous St. Peter's Basilica in Rome had just been completed, and artists such as Michelangelo were creating art and statuary that still hold people in awe today.

At the same time, for those who looked closely, there were signs of cracks in the façade of the glorious Renaissance church in Europe. Early on, the perennial struggle between the nobility and the hierarchy became violent when the King of England Henry II ordered Thomas Becket, the Archbishop of Canterbury, murdered in 1170 in a dispute about the rights and privileges of the church. Church and State continued to vie for power and wealth, and many nobles wanted to escape from the control of local bishops and the pope in Rome.

The papacy was often an embarrassment. In 1302 Pope Boniface VIII arrogantly decreed that he had highest authority over the whole world and that all had to obey him in order to be saved. In the fourteenth century the papacy had been moved to Avignon, France. Eventually, there were two popes and then three popes at a time vying for power. The papacy hit its nadir when Alexander VI, a member of the corrupt Borgia family and the father of a number of illegitimate children, was elected Pope in 1492.[18] Great wealth passed hands as bishops had to purchase their positions, and nobles had to pay money to the church in Rome in the form of taxes and favors.

The Black Death profoundly added to the deterioration of the church's situation. This fourteenth-century plague killed off perhaps half of the population of Europe and crippled its institutions, especially clergy who ministered to the sick and dying. Playing upon the peoples' terrors of death, the church offered the opportunity to buy **indulgences**, which granted full or partial remission from punishment for sin and were purported to save penitents from final damnation and hell.

There were also new challenges to orthodoxy. The brilliant Oxford professor John Wycliffe (1331–1384) proposed that the ordinary people should have access to the Bible in their own language and challenged literal interpretations of the real presence of Christ in the Eucharist. Wycliffe managed to escape the Inquisition; however, his Bohemian (Czech) follower Jan Hus went to Rome in 1415 thinking he was summoned for a dialogue—he ended up being burned at the stake for heresy.

It should be noted that there were many at the time who wanted to take their Christianity more seriously and return to the principles and practices of the early church. Often they were not well received by the official church. Around 1170 Peter Waldo renounced his wealth and began to live a life of poverty and prayer in imitation of Jesus. He had some of the gospels translated into French and began preaching from town to town. His followers, both men and women, formed the *Waldensians* and traveled about preaching the gospel message to the illiterate masses. Since these lay people were not authorized by the church, they were forbidden to preach. When they continued, they were condemned and excommunicated.

Between 1150 and 1250 a group called the *Cathars* flourished in southern France and Italy. Seeing themselves caught in the struggle between the forces of good and evil, they chose a radical gospel lifestyle, some rejecting the material world, sex, and marriage. They ignored all sacraments except a spiritual baptism, which was a type of born-again experience. In 1208 Pope Innocent III unleashed the Inquisition and a crusade against the Cathars. After their last defeat, their bishop and hundreds of Cathars were burned alive.[19]

Those who pursued an intense prayer life were also punished. Women who claimed to have visions and mystical experiences were often condemned and burned as witches. The accomplished mystic Meister Eckhart (1260–1327) dedicated himself to prayer and proposed that God could be experienced on one's own, without benefit of church or clergy. His extensive mystical writings were condemned. Savonarola (1452–1498) was a Dominican friar who claimed to have visions of the end to the world, with judgment coming upon Florence and the church for its excesses and corruption. He was an extremely popular preacher and loudly condemned the riches of the church. The infamous Pope Alexander VI excommunicated Savonarola. Later, the friar was hung and burned in the public square in Florence.

Martin Luther – Fear of the Inquisition, disillusionment with the corruption of the church hierarchy, bitterness from the excommunications and executions, and the experience of chaos and death from the plague and other factors provided kindling for reform. All that was needed was for someone to drop the match.

An unlikely person ignited this greatest conflagration in Christian history, now called the Protestant Reformation. His name was Martin Luther (1483–1546), a bright, young Augustinian monk, who taught Scripture at the University of Wittenberg in Germany. Martin's scrupulosity about his sinfulness and his preoccupation with guilt and fear of hell gave him much spiritual anguish. During an intense personal conversion experience, Martin read Paul's Epistle to the Romans: "The one who is righteous by faith shall live" (Romans 1:17). He suddenly realized that he could do nothing to save himself other than turn in faith to God, who would cover his sins with grace and save him. From then on, Scripture alone would be his guide, and faith alone would save him. Luther would make belief in "Scripture alone" and "faith alone" foundational to his later theology.

Luther's belief in the power of faith alone to save led him to be critical of the sale of indulgences promoted locally by a Dominican named John Tetzel. At the time, Luther had no quibble with indulgences themselves, since they could take away punishment due to sin. But paying money for these indulgences was unacceptable to him because it seemed like buying salvation rather than gaining it through faith. Luther requested a debate about these and other religious matters, and in the fall of 1517 he was said to have posted his now famous *Ninety-Five Theses* on the castle church door, which was the standard manner of calling for a debate at the university.

Luther must have been amazed when his theses criticizing indulgences were printed, circulated, and celebrated by many as a challenge to the pope's authority. This young monk suddenly became the public image for all the monarchs and nobles who wanted release from the pope's authority, as well as for all those who were disgusted by the corruption in the church and alienated by the many horrors brought on by the Inquisition.

>> *See Martin Luther at the Diet of Worms. Why do you think Luther answered the way he did when asked to recant his statements?*
http://www.youtube.com/watch?v=r5P7QkHCfaI&feature=related

At first, the pope ignored the criticisms of this seemingly insignificant monk, but when he learned of the publicity Luther was getting, the pope called him to Rome. The summons was definitely not friendly, calling Luther "a leper and loathsome fellow...a dog and son of a bitch, born to bite and snap at the sky with his doggish mouth."[20] Afraid of execution, Luther declined the invitation and, being a bombastic person, moved to a more hostile and extreme position. A cardinal was sent from Rome, but got nowhere with Luther. Rome became distracted with political matters in the Holy Roman Empire, and Luther had several years to publish sermons, pamphlets, and treatises radically challenging the legitimacy of the papacy, clericalism, religious life, church hierarchy, and all the sacraments other than baptism and Eucharist. In time, many of his propositions were condemned, and Luther answered by calling the pope the Antichrist. In 1520 Luther burned the papal bull that condemned him.

The young monk was called before the emperor and church officials and ordered to recant. Luther refused and made his famous statement: "Here I stand." Martin's local political leader, Frederick III, Elector of Saxony, fearing that the young monk would be executed, feigned Luther's kidnapping and hid him in a castle, where Luther used his time to complete his brilliant translation of the Bible into German.

Luther's movement, supported by many of the secular rulers, spread rapidly through northern Germany. Monasteries and convents were sacked, and many bishops were stripped of their wealth and power. Luther married a former nun and settled into what had been his former monastery, where he continued to write and hold council. Several serious efforts were extended for reconciliation with the church, but these failed, and religious wars broke out after Luther's death.[21] In 1555 the divisions were finalized, and the faith of the local leader determined whether an area would be Protestant or Catholic. Luther had never intended to start a new church, but his followers now referred to themselves as Lutherans.

The reform began to spread under other leaders. In England, King Henry VIII (1491–1547) broke with the pope over the refusal of a marriage annulment that would allow him to take a second wife. Henry rebelled, declared that he was head of the church in England, and confiscated vast amounts of properties owned by the church. Under King Edward VI (1537–1553), England became more Protestant, but reverted to Catholicism under the reign of Mary Tudor (1516–1558), and then settled into a moderate Protestantism under Queen Elizabeth I (1533–1603), a form now known as **Anglicanism**. The Anglican churches have maintained much of traditional creedal doctrine, liturgy, and clergy offices. They have also embraced some Protestant practices by rejecting the papacy, allowing the clergy to marry, and developing their own *Book of Common Prayer*. In the United States these churches would be referred to as Episcopalian.

>> *Watch* **Reformation Overview: Zwingli** *and consider the emphasis Zwingli placed on the words of the Bible over the doctrines of the Roman Catholic Church:* http://www.youtube.com/watch?v=FI2oFvEUfXE

The Reform was brought to Switzerland by Ulrich Zwingli (1484–1531), a former priest who taught that the Bible was to be the only Christian source and that statues, altars, relics, religious life, and all sacraments other than baptism and Eucharist should be abolished. A visit to his church in Zurich today still strikes one with its stark bareness—only a large Bible is given prominence. With regard to the Eucharist, Zwingli was more radical than Luther. He denied the real presence of Jesus in the host and chalice. He also opposed the **Anabaptists**, who forbade infant baptism, and drowned many of such "heretics" in Lake Zurich. Zwingli was warlike in his convictions and died in battle with Catholics. His body was quartered and he was burned as a heretic, but his movement survived and spread through Switzerland.[22]

>> *Watch the video* **Reformation Overview: Calvin** *to learn more about the development of Calvinism and the writings of the reformed Christian manual Institutes of the Christian Religion:* http://www.youtube.com/watch?v=9VvIQP2gg8g&feature=related

John Calvin – Another key reformer was John Calvin (1509–1564), who was drawn to the Protestant reform in 1533 in France. When Calvin witnessed reformers being executed, he moved to Basel in Switzerland

and later to Geneva. At first, he was too strict for the populace and they threw him out, only to invite him back later to establish Geneva as the ideal Protestant city. Calvin then wrote his *Institutes of the Christian Religion*, a brilliant handbook, and continued to rule Geneva with a firm hand. Those who would not go along with the strict morals of **Calvinism** were punished, some beheaded, or even burned at the stake. Calvin took a radical position against the real presence of Jesus in the Eucharist and set up his own church structure of presbyters (elders).[23] In his *Institutes of the Christian Religion* (1536), Calvin rewrote the Christian doctrinal system. His most controversial teaching was predestination, the belief that God predetermines those who go to heaven and those who go to hell. After Calvin's death, this teaching became a hotly debated pastoral question in his churches. Calvin's version of Protestantism spread through much of Europe to Germany, Poland, Bohemia, Hungary, Moravia, France, Netherlands, and Belgium. John Knox (1510–1572), who was trained by Calvin, spread Calvinism to Scotland, where it developed into the Presbyterian Church of Scotland. Calvinists surfaced in England and were called **Puritans** because of their efforts to purify England of anything "popish." When they came under persecution, many of them fled to the New World, where they established the Massachusetts Bay Colony. Today they are known in the United States as Presbyterians, and it is thought that much of the work ethic and the linking of religion with government in the United States are derived from this church's teaching on the virtue of hard work.

Protestant Groups – Other churches evolved out of Protestantism. The **Baptist** tradition started in 1609 in Amsterdam with John Smyth, who opposed infant baptism, opting for a believer's baptism and the autonomy of local churches. Baptists divided into many denominations that differ in their beliefs.

George Fox started the Quakers, or the Religious Society of Friends, in 1646 in England. Quakers believe in an inner light that represents a sense of God, and they meet to experience this light, with no need for ministers. They have been strong advocates for nonviolence and were early opponents of slavery.

>> *Identify the way the Quaker meeting and speaking to the group differs from other Christian worship services in* An Introduction to Quakers:
http://www.youtube.com/watch?v=q-rdlmcwTFw

John Wesley, an indefatigable preacher in the early eighteenth century, who began a reform movement in the Anglican Church of England, started the Methodists. Methodism stresses Bible study, the "methodical" and disciplined practice of the Christian life, simple liturgy, and service of the needy. Methodism moved to the United States in the frontier days. Its firebrand preachers were a large draw in the rural areas.

The **Mennonites** are an offshoot from the Anabaptist group in Switzerland. They were founded by Menno Simons and are dedicated to living out the message of the Christian gospel simply and to nonviolence.

>> *Consider how the Amish make their daily life a commitment to Christ.* Amish: A People of Preservation:
http://www.youtube.com/watch?v=qXmrxWmbHCU&feature=related

The **Amish** are a reform of the **Mennonite** movement and were founded in Switzerland in 1693 by Jakob Ammann. They opt for a simple Christian life, wear simple uniform dress, and live further removed from the modern world and its conveniences than do the Mennonites.

Pentecostalism is an international movement that began in the early twentieth century and is rapidly growing in the United States, Central America, and South America. Pentecostals emphasize the gifts of the Holy Spirit, including speaking in tongues, prophecy, and healing.

THE ROMAN CATHOLIC REFORMATION

The Protestant movement understandably shook Roman Catholicism. Some church reform had begun earlier and was now seriously intensified in what some refer to as the Counter Reformation. The training of priests was restructured and improved while new groups formed to revitalize the church's educational and mission activity. The Society of Jesus, founded by Ignatius of Loyola in 1534, became the vanguard for the papacy and made major contributions to the renewal of the church. Starting as a small community of men in Spain, the Society (Jesuits) grew quickly and spread through Europe, Asia, and the New World as missionaries, teachers, pastors, and theologians. Peter Canisius, a Jesuit and a major reformer in Germany, established colleges and published a catechism. The Capuchin friars are a sixteenth-century reform of the Franciscan friars; the great mystical writers John of the Cross and Teresa of Avila reformed the Carmelites. New orders of Sisters were established and contributed mightily to the reform of Catholicism, providing hospitals, schools, and many other services.

The papacy was slow to call a Council for reform, seemingly fearful that such a Council would dilute papal authority. Finally, in 1545 the Council of Trent assembled and met sporadically until it finally finished its work in 1563, four decades after Luther had challenged the church.

>> Listen to the history of the Council of Trent and determine
the reasons why the council took eighteen years to finish its work:
http://www.youtube.com/watch?v=NZXq10_zdl8

The eventual decrees of Trent were clear and definitive and would set the church on a course for the next four hundred years. The objections of the Protestants were addressed, and there was an adamant hardening in the areas that had been abandoned by the reformers. It was decreed that **justification** (saving of one's soul) was not only by faith, but also by hope and charity, the latter expressed through good works. **Original sin** (the sin inherited from Adam and Eve) wounded human nature, but did not totally corrupt it. All seven sacraments were defined, and firm rules were delineated for their celebration. The **Mass** (the ritual of Eucharist) was clearly and definitively organized. The real presence in Eucharist was strongly affirmed, using the medieval term of **transubstantiation**: the matter of bread and wine remains the same but the substance changes into the body and blood of Christ. The papacy was defined as "divinely instituted," and careful guidelines determined the selection of cardinals. Bishops were required to live in their dioceses and be pastoral leaders, and the purchase of their offices was forbidden. The ordained priesthood was affirmed, and celibacy required and enforced. Seminaries were strengthened through sound recruiting and strict training. Both Scripture and tradition were deemed necessary to constitute the faith, and Scripture could only be interpreted by the church, not by private individuals. The Douay-Rheims translation of the Bible was completed in France to provide a Catholic Bible to the faithful.[24]

The Catholic reform was effective; Catholicism was able to regain ground in Germany and remain strong in France, the Netherlands, and Poland. Catholic missions overseas were successful. The papacy remained free of scandal, and a strong system of dioceses was established throughout Europe and eventually in the mission areas, led by carefully selected bishops. Catholic universities and theology flourished.

The Catholic reform moved the church into a defensive, fortress mentality that caused it to resist later developments in Europe, such as the birth of modern science with the cosmology of Galileo (1564–1642), the rise of rationalism in the Enlightenment (1650–1800), the development of democracy, and the emphases on freedom and nationalism in the eighteenth and nineteenth centuries. The French Revolution and the Cultural Revolution in Germany brought much persecution to the church, which further reinforced this defensive position.

THE ENLIGHTENMENT

The Enlightenment of the seventeenth and eighteenth centuries challenged all Christian churches. "Reason alone" challenged "faith alone." René Descartes set the agenda with his famous statement: "I think, therefore I am." Many of the educated no longer looked to their Christian religion for truth, but rather to philosophers such as Francis Bacon, René Descartes, John Locke, and David Hume. Others believed that truth and progress would come from science and turned to thinkers like Galileo Galilei, Isaac Newton, and Johannes Kepler. Even those still disposed to religion were discerning that religious truths such as the existence of God, ethical norms, and the afterlife could be discovered by reason alone. The notion of *revelation* was challenged, and a natural religion would ultimately be proposed that would deny the supernatural, miracles, and the mystical.

During this period, a movement called **Deism** emerged that held that the only worthwhile Christian beliefs were those provable by reason. This led to a rejection of belief in the divine inspiration of Scripture and a discrediting of anything supernatural or miraculous in the Bible. Deists posited a distant "Sky God," who did not interfere with human affairs or the laws of nature. The Deists also set the divinity of Jesus and belief in his miracles aside.[25] Some of the founding fathers, such as Benjamin Franklin, Thomas Jefferson, Thomas Paine, and Alexander Hamilton, may have been influenced by Deism. Thomas Jefferson even took it upon himself to write his own version of the gospels, leaving out any reference to the supernatural.

THE EVANGELICAL MOVEMENT

In the eighteenth century there were also movements that came to be known as evangelical Christianity. This movement is rooted in the Bible and emphasizes the experience of emotion and personal conversion. Within this movement might be included the Methodists, founded by

John Wesley (1703–1791), who carried the gospel reform message of conversion and service on horseback throughout England. The movement spread to America.

The evangelical movement was given great impetus in America during The First Great Awakening in the eighteenth century. One of its leading figures was Jonathan Edwards (1703–1758), a powerful and fiery preacher. His sermons, as well as those of George Whitefield (1714–1770), brought many converts to the Protestant denominations. There was another Great Awakening in America in the nineteenth century, carried on by revivalists such as Charles Finney (1792–1875). In the twentieth century a strong evangelical movement stirred many Americans, led by powerful preachers such as Billy Graham (b. 1918). The evangelical movement remains both a strong religious and political force in the United States.

THE MODERN PERIOD

Rationalism and science would further challenge the Christian churches and congregations in the nineteenth century. Modern biblical criticism was an effort to examine Scripture scientifically, using linguistics and cultural context. The movement, which often viewed Scripture in terms of literary forms such as myth, became quite threatening to those who had accepted Scripture literally.

Charles Darwin's theory of evolution also severely challenged the perennial belief in creation by God and would be troubling to churches even into modern times. The rise of atheism and Communism in the nineteenth century brought serious challenges and often persecution to Christians, as we shall see in more detail later.

Early in the twentieth century, the consciences of Christians were challenged by the Social Gospel movement led by Walter Rauschenbusch (1861–1918) as well as by Catholic social teachings, which advocated for social justice and supported labor unions. Christians were being called to serve the poor and underprivileged. Another movement that caused division within Protestantism during the twentieth century was Neo-Orthodoxy. Led by the brilliant Scripture scholar Karl Barth (1886–1968), Neo-Orthodoxy reminded Christians of the radical demands in the gospels as well as their responsibility to resist evil. This movement led many in Germany to resist Nazism, including Pastor Dietrich Bonhoeffer (1906–1945), who was executed for his involvement in a plot to assassinate Hitler.

Today, many Christian churches still struggle with the progressive views of theologians and biblical scholars who often challenge the divine inspiration of Scripture, traditional beliefs about creation, the divinity of Jesus, the ordination of women and gays, cohabitation, and many other issues. Medical science now offers procedures such as abortion, sex changes, assisted suicide, cloning, and stem cell research. These issues have also presented serious challenges to the moral standards of the Christian churches and congregations.

THE CHRISTIAN SCRIPTURES

The final collection of Christian Scriptures was determined only after centuries of discernment and controversy. One of the first conflicts was with Marcion of Pontus (85–160 BCE). Marcion was a Christian bishop, one of the early heretics, who started his own assembly called the Marcionites. Marcion did not accept the legitimacy of the Jewish Creator God. For him, Jesus was the revelation of the redeemer God, a spiritual God, whose Son was not "of the flesh." Marcion proposed a Christianity that rejected the created world, materiality, sex, and all things of the flesh.[26] He insisted that the Hebrew Scriptures with its Creation God be excluded from the Christian collection. His views spread through some of the communities around the Mediterranean and lasted for centuries. Mainstream Christianity rejected Marcion's views, insisting that the Jewish God was the same as the Christian God and that Jesus was his Son in the flesh, the fulfillment of the Hebrew prophecies. The Hebrew Scriptures were therefore to be part of the Christian Scriptures. Marcion was excommunicated.

Other conflicts arose with the so-called Gnostic communities. **Gnosticism** is a general term applied to those who claimed to have some mystical knowledge and esoteric insights that others did not have. Christian Gnostics usually rejected the material world and, therefore, could not accept the *incarnation*, the belief that God could come "in the flesh" as Jesus Christ. Gnostics recognized the value of the feminine and gave women more authority in their communities. These groups developed their own gospels and other Scriptures, generally quite different from those of the church. These writings were rejected by official Christianity and were believed destroyed until a huge and significant collection of them was discovered near Nag Hammadi, Egypt, in 1945. Since then, these writings have been intensely studied, and scholars are divided with regard to their value in the Christian tradition.

The Christian Bible – The Canon, or official collection of Scriptures in the Christian Bible, evolved gradually. In the second century the Apostolic Fathers, Greek authors of early Christian works whose writings were considered to be in continuity with the apostles, seemed to be working with a certain collection of Scriptures. Great preachers, such as Ambrose (340–397), the bishop of Milan, and Augustine (354–430), the bishop of Hippo in Africa, drew their sermons from the Bible. The Apostolic Constitutions (350–380) list the specific books that are in the canon. The collection was likely finalized in 382. Around that same time Jerome (345–420), a Christian biblical scholar, was commissioned by the pope to prepare a definitive Latin translation of the Bible. The translation was brilliantly done and stands even today as having great significance in Bible studies.[27]

The Scriptural Texts – The Christian Bible consists of the Old Testament (the Hebrew Scriptures) and the New Testament. The New Testament is a library of twenty-seven books and includes the four gospels, the

Acts of the Apostles, early letters written by leaders in the communities such as Paul, and the Book of Revelation. Christians reflect on both the Old and New Testaments to draw conclusions regarding their faith, and both Testaments are used in religious teachings and rituals.

The Gospels – The centerpiece of the New Testament is the collection of four gospels. The word *gospel* means *good news*, so the title reflects the exciting and joyful witness of the early Christians, who followed Jesus Christ as their savior. They believed that if they lived good lives of love and compassion they would live on after death. The gospel of John proclaims Jesus' promise: "Truly, truly, I say to you, the one who hears my word, and believes the One who sent me, has eternal life, and does not come into judgment, but has passed out of death into life" (John 5:24).

The gospels were originally written in Greek and were named after various apostles and companions of the apostles. The real authors seem to be anonymous Christian writers, who recorded the beliefs of the communities in which they lived. The gospels began as oral traditions, which were in time written down. Many biblical scholars propose that underlying these stories is the post-Resurrection faith in the communities. From this perspective, this faith both colors the stories and interprets who Jesus is, as well as what he says and does. In other words, these scholars suggest that what we have in the gospels are not historical accounts or actual verbatim dialogues. Rather, they are memories of what Jesus said and did transformed by the later faith of the believers. For example, in the four *passion stories* the foundation is the memories of the horrible experience of seeing Jesus betrayed and accused, scourged, tried, and executed by crucifixion. In the gospels, these memories are transformed into four different plays that teach about faith, courage, forgiveness, and love. The gospels, then, are not histories and biographies, but rather faith stories that teach the fundamentals of who Jesus was and what he taught.[28]

Mark's gospel was the first gospel, written in 70 CE, possibly in Rome or Syria. It seems to come from a non-Jewish community that was being persecuted. The gospel opens with Jesus as a young man about to begin his mission. At first, he was received with much enthusiasm, but as the story proceeds he began to make enemies with the Jewish and Roman leaders. Eventually, Jesus was condemned and crucified. The story of the Resurrection is early and quite simple. On the Sunday morning after Jesus' death, women followers of Jesus find the tomb empty and are told that Jesus has been raised; they are directed to tell Peter and the others. The women leave the tomb terrified and tell no one.

Matthew's gospel was written around 80 CE, probably in Antioch, Syria, a predominantly Jewish-Christian community. It stresses that Jesus is the fulfillment of the hopes of the Jewish people and goes into great detail about Jesus' birth, his ministry, sermons, and miracles. Much of this gospel is taken from Mark's gospel, but added are many quotations from a source known as *The Sayings of Jesus*. The events surrounding the Resurrection are told here in more detail. When the women are told that Jesus had been raised and to tell the others, they do that with great excitement. On the way, they encounter the risen Jesus and worship him. The eleven apostles later encounter the risen Jesus on a mountain in Galilee and are commissioned to spread his message to the world.

Luke's gospel was written sometime in the 80s near Syria or Greece. The author is a master theologian and an accomplished writer in Greek. He has much regard for Mary the mother of Jesus. His account of the passion is unique, stressing Jesus' forgiveness of his enemies. Luke's post-Resurrection stories are vivid, stressing the breaking of the bread when Jesus dines with two followers in Emmaus and describing his final words to the eleven in Jerusalem.

The gospel of John is unique among the gospels. It seems to have been begun in Ephesus around 90 CE and, after a number of editions, was finished around 100 CE. John's gospel presents advanced thinking about Jesus, his incarnation, and the miraculous signs he worked. This gospel uses profound symbols of water and wine, light, bread, vines and branches. The symbols would later be the basis for a "sacramental" theology developed by Augustine in the fourth century. Here, sacraments are described as sacred symbols, which put the devout in touch with God and his divine grace. The miracles and discourses reveal a divine Jesus who promises to send his Spirit. He reigns over enemies throughout the passion, and the Resurrection stories are intimate, as in the account of when the risen Jesus cooks breakfast for his apostles on the beach.

The Acts of the Apostles is a second volume by the author of Luke's gospel. It is a history of the early community told with poetry and theology woven into the narrative. It includes stories of the risen Jesus appearing to his disciples, promising his power and support and then ascending out of sight. Acts gives an account of the first Pentecost and stories of the faith and lifestyle of the early Christians; it also shows how they suffer in order to carry out their mission. Several accounts are given of Paul's conversion and his journeys to expand Christianity, as well as his being sent to Rome to face death.

>> *As you watch the video* **St. Paul the Apostle, Rome, Italy** *consider Paul's story and the impact his conversion had on the establishment of Christianity:* *http://www.youtube.com/watch?v=yeEjCgY3-r4*

The New Testament includes the Letters of Paul. As mentioned earlier, Paul was a Pharisee who persecuted Christians. Struck down and converted by an experience of the risen Jesus, Paul undertook a mission to bring Christianity to non-Jews, the Gentiles. He traveled thousands of miles on foot and by boat, and suffered beatings, imprisonment, shipwreck, and ultimately execution in Rome. Paul follows Jesus' example in inviting women to be leaders in his mission—Lydia, Priscilla, Phoebe, and Junia. In his letters or epistles to the early communities, Paul developed a profound theology about Jesus and his message, his divinity, the meaning of the cross, and even Christ's cosmic significance, which has been extraordinarily influential among great Christian leaders, including Augustine, Aquinas, Luther, and Pope John Paul II. This theology would leave a powerful stamp on Christianity.[29] Many scholars maintain that only some of these letters were actually written by Paul: 1 Thessalonians, Philippians, 1 & 2 Corinthians, Galatians, and Romans. Paul's disciples probably wrote the other epistles.

In addition to Paul, there are other letters by Christian authors in the New Testament: Peter, Hebrews, and Timothy. The final section of the New Testament, the Book of Revelation, is a highly poetic and symbolic reflection on the endtime, also known as the Day of Judgment, as well as a reflection on the early Christian struggle with the Roman Empire.

CHRISTIAN RITUALS AND FEASTS

From its inception, Christianity, like many other religions, had need for rituals and symbols that would put them in touch with the sacred. The Christian "sacramental principle" is that the hidden mystery of God's presence can be revealed and experienced through symbols and rituals. Sacraments are "sacred symbols and rituals through which we can be uniquely put in touch with the spiritual, the very power of God."[30] Each of the traditional seven sacraments has its own history of development, and each continues to be reinterpreted and practiced differently today. After the Reformation, the Catholic Church maintained that there were seven sacraments, while the reformers held for only two: baptism and Eucharist.

Baptism – Borrowing from the Jewish water purification rites, and in commemoration of Jesus' own baptism in the Jordan by John the Baptist, the early Christians used **baptism** with water as their own initiation rite. Usually it involved submerging the candidates in water, which symbolized going down into the death of Jesus. Rising from the water symbolized a rebirth into the Resurrection of Jesus. Baptism was viewed as both a cleansing from sin and as an entrance into the community of Jesus Christ. Adult baptism was the norm in the early church, but eventually infant baptism became the custom. There has always been a debate among the churches on this issue. Catholics, the Eastern churches, and many Protestant churches still baptize infants. Baptist and many evangelical churches only baptize those who are adult enough to choose being "reborn."

Eucharist – Eucharist is a Greek word that means *thanksgiving*. From the earliest times, Christians gathered in homes for "the breaking of the bread," to thank God for their Christian faith and salvation, to commemorate Jesus' last supper with his disciples, and to follow his request to "Do this in my memory" (Luke 22:19). Paul speaks of this meal as early as 50 CE when he writes to the Corinthians that "as often as you as you eat this bread and drink this cup, you proclaim the death of the Lord until he comes again"(1 Corinthians 11:26).

Debates over the "realness" of the real presence of Jesus in Eucharist began as early as the eighth century. Some taught that Eucharist was literally the body and blood of Christ, and others believed that Jesus was only symbolically present. This debate intensified during the Reformation, with Luther remaining a literalist, but Zwingli and Calvin insisting that Jesus was present symbolically. The Catholic Church took a strong position on the real presence and used the word "transubstantiation," meaning that while the *accidents* (color, taste, etc.) remained the same, the very substance of the bread and wine was changed into the body and blood of Christ. Today many Christians have moved away from either literal or purely symbolic interpretations and describe the real presence as the "spiritual" presence of the Risen Lord. Catholics, both West and East, as well as many traditional Protestant churches celebrate Eucharist every Sunday. Catholics call their service *Mass*, which is a formal ceremony presided over by a priest. Mass consists of formal prayers, readings from the Scriptures, a sermon, offering of the gifts of bread and wine, a solemn consecration of these gifts into Eucharist, and then the distribution of communion to the faithful. Other churches hold services for worship and preaching the Word, but offer communion less often, usually accompanying the bread with grape juice rather than wine.

Sacraments – As mentioned earlier, the Catholic Church still maintains seven sacraments. The church speaks of **Confirmation** as the sacrament that completes baptism, when the person receives the special strength of the Holy Spirit. **Penance** or reconciliation is a ritual for requesting forgiveness and is celebrated with a priest. The **Anointing of the Sick** is a sacrament of healing, offered to those who are in danger of death or who are in the dying process. **Marriage** is a sacrament witnessed by a priest, wherein the solemn marriage vows are exchanged. This sacrament is viewed as a symbol of a commitment to permanency and fidelity. **Ordination** is the ritual whereby men are ordained to the priesthood or to the deaconate.

The Christian Year – The Christian year centers around two events in the life of Jesus Christ: his birth and his Resurrection. **Christmas**, the celebration of Jesus' birth, is preceded by the four-week preparation of Advent (from the Latin word *adventus*, which means coming).

Christmas is always celebrated on December 25, so Advent begins the fourth Sunday prior to that date and is a time of prayerful waiting and preparation for the coming of Jesus. In church and in many homes an Advent wreath with four candles becomes the center of special prayers.

The celebration of Christmas is special for Christians and is usually marked by a church service and the assembling of Christmas cribs in churches and homes. There is often an effort to tie in the gift giving and decoration that goes on in the culture with the joy of commemorating the birth of the savior. God's gift of Jesus prompts Christians to offer gifts to others.

The most important feast of the Christian year is **Easter**, the celebration of Jesus' Resurrection three days after his crucifixion. This feast is anticipated by Catholics during a six-week period of fasting, penance, and preparation. The forty days of fasting honors the forty days that Jesus spent fasting in the desert before he began his ministry. The week leading up to Easter is often called Holy Week, which opens with Palm Sunday, the celebration of Jesus' triumphant entrance into Jerusalem before his arrest and death. On Thursday, services are held to commemorate Jesus' last supper with his disciples. Good Friday is a solemn day with services commemorating Jesus' suffering and death on a cross. Sunday is the day to celebrate with joy and thanksgiving the Resurrection of Jesus. Many Christians celebrate Jesus' ascension into heaven forty days after Easter, and Pentecost, the birth of the church, fifty days after Easter.

The various Christian denominations celebrate the Christian calendar in varying degrees. Roman and Eastern Catholics and most mainline Protestant churches celebrate the liturgical year in its entirety. The Orthodox churches follow a different calendar, so the feasts fall on dates that are somewhat later.

Catholics often celebrate days dedicated to the saints of their church or feasts honoring Mary, the mother of Jesus. Catholics still revere the Blessed Virgin as the Mother of God, pay her honor with special prayers such as the rosary, celebrate her birth without original sin (the Immaculate Conception), and acknowledge her entrance, both body and soul, into heaven after death—a feast day known as the Assumption.

IMPORTANT MODERN DEVELOPMENTS

There have been many important developments in Christianity. The three that will be discussed here are: ecumenism, the Second Vatican Council, and liberation theology.

Ecumenism – In the United States many Christians were isolationists before World War II, but from the war's outset most staunchly supported the American war effort. After the war, many Christians improved their education through the GI bill, moved up the social ladder, and established their churches in the suburbs. Ethnic ghettos became a thing of the past. Protestants and Catholics collaborated, and there was a marked increase in intermarriage. At the same time, a new evangelism caught on, led by preachers such as Billy Graham, and eventually moving on to the new medium, television. The World Council of Churches was established in 1948 with its goal of Christian unity and cooperation. The Council convened in the United States in 1954, with most churches, other than Roman Catholic, participating. The Second Vatican Council encouraged dialogue and mutual liturgical services among the churches. During the 1970s and '80s, such dialogues and services were quite popular, but currently interest has lagged. Many churches are drawing within their own traditions in a more conservative mode.

The Second Vatican Council – The turbulent 1960s began with the landmark election of the first Catholic president of the United States, John F. Kennedy. In 1962 the **Second Vatican Council**, otherwise known as Vatican II, opened a series of meetings that would deeply affect Catholics as well as Protestants in the United States and worldwide. This Council, consisting of more than two thousand bishops and cardinals, along with many Protestant observers who were valuable contributors, lasted for a period of three years. The Council reformed Catholicism's liturgy and made commitments to Scripture, peace and justice in the modern world, religious freedom, and open dialogue with other churches and religions. The Council spoke of the church as "the people of God," a "pilgrim people," and tried to heal the wounds of the past. Since that time there have been many overtures for reunion with the Eastern churches. Christian church attendance is significantly down in Europe and dropping in the United States in the mainstream churches. In the United States the evangelical mega-churches have gained popularity. These churches usually have few doctrinal commitments and conduct lively services, often using media and contemporary music, and address social justice and more personal issues that are relevant to young people. The Anglican and Episcopal churches today are seriously divided over the ordination of openly homosexual bishops. Many Episcopalians in the United States are threatening or actually leaving their communities because of this issue.

Since the 1960s there has been a significant exodus of Catholic priests and nuns, and religious communities have fewer applicants. The Vatican has taken a strong stand against the ordination of women, but there are growing movements that favor such ordinations, such as the Women-Church movement, the Women's Ordination Conference, and Roman Catholic Womenpriests.

The Catholic Church has been seriously damaged by the scandal of priests and even bishops accused of sexually abusing children in Canada, the United States, and Europe. Numerous reforms have been put in place, and many dioceses have paid large sums to victims. In both Catholic and Protestant churches in the United States there is a growing division between conservatives and liberals, traditionalists and progressives. There seems to be little common ground among these groups, and there is reason to believe that there will be serious divisions in the various Christian churches in the near future.

Liberation Theology – Vatican II was a watershed for Catholicism in South America. The Council directed the church's attention to the poor and to social justice. The bishops of the southern hemisphere gathered in 1968 and pledged themselves to serve the poor.

Liberation theologians interpreted the gospels to energize the poor toward freedom. Base Christian communities were established for prayer, Bible study, and consciousness-raising. Many of the clergy struggled against the oppression of the poor and were killed by the government. Two Catholic prelates were outstanding in their solidarity with the poor: Archbishop Helder Camara of Recife in Brazil and Archbishop Oscar Romero of El Salvador. Romero was murdered by the government for his exposure of its corruption and for daring to order from the pulpit that the military stop their oppression of the poor. The government of El Salvador also murdered four American women missionaries as well as six Jesuits at the University of Central America along with their housekeeper and her daughter.

GLOBAL CHRISTIANITY

Christians believe that Jesus commissioned his disciples to "Go, therefore and make disciples of all nations" (Matthew 28:19). Today, Christianity is the largest of the world religions and is practiced on all the continents, with the greater percentage of growth found outside of Europe, particularly in Africa, South America, and Asia. The major shift today in these continents outside of Europe is that they are moving from having an imported European Christianity flavor to developing a truly "global Christianity" that embraces and serves local cultures.

India – Christianity first came to India from Syria in the first century CE, then from Portuguese explorers and Jesuit missionaries in the 1600s, and finally in the eighteenth and nineteenth centuries with the advent of the British colonials. Mother Teresa became an excellent example of the modern Christian missionary, one who respects and serves the Indian religions and culture. Though there are 50 million Christians (Catholics being the larger group) in India, they still remain a small minority, yet one known for energy and growth.

Africa – As mentioned earlier, the early Christian faith spread through northern Africa. Great Christian centers in Alexandria and Carthage were established, producing such giants as Augustine of Hippo and Cyprian of Carthage. It was in Africa that the monastic movement was started by the Fathers and Mothers of the Desert. In the fifteenth century the Portuguese explorers brought the next wave of Christianity to the Congo area of Africa. The missions spread to West Africa during the infamous slave trade.[31] Some missionaries ignored this horror, but others created safe havens for the natives and resisted the slave trade. The Quakers opposed slavery from the outset, and Christian resistance grew until 1807, when the slave trade was outlawed in Britain.

The white settlers in South Africa, the Afrikaners, looked down upon the natives and worshiped in the Dutch Reform church, while the native

South Africans set up their own Christian churches. Some English missionaries fought for the dignity of the tribal people and the end of forced labor, but the colonials resented their efforts. In recent times the churches became quite involved in the successful Black struggle to overthrow apartheid, led by such outstanding church leaders as Bishop Desmond Tutu, Trevor Huddleston, Allan Boesak, and Bishop Denis Hurley.

>> *See Desmond Tutu on leadership:*
http://www.youtube.com/watch?v=IrCeVwwuOXc

The Catholic Church and the Pentecostal churches have seen enormous growth in Africa. There are presently more than 494 million Christians in Africa.[32] It is projected that by the end of this century, Christians in Africa will become more numerous than Christians in any other single continent.[33]

Canada – In Canada, Catholicism was historically quite vibrant in French Quebec, but it now seems to be weakening under the pressure of secularism. A "quiet revolution" has taken place wherein the church has lost much of its control over labor and education, and church attendance has dropped significantly.[34] Catholics throughout the rest of Canada often find themselves feeling the same split between conservatives and progressives and experience more commonality with their Protestant acquaintances. Overall, there are more than 283 million Christians in North America, with more than 21 million Christians in Canada.[35]

Central and South America – The Catholic Church was established in Central and South America by Spain in a strict traditional style. This was a highly devotional church, given to dedication to the Virgin Mary, local saints such as Rose of Lima, and especially to Our Lady of Guadalupe, who was said to have appeared to Juan Diego, a poor Indian, in 1531 near Mexico City.

In the 1950s and '60s many Catholic missionaries from Europe and the United States flooded into Central and South America to rejuvenate the church. By 1960, 35 percent of the world's Catholics lived in Central and South America, more than in Europe. During the revolutions, the church's support began to shift to the poor, and as a result many missionaries, priests, and nuns were deported, tortured, or killed. An archbishop, Oscar Romero, was assassinated while celebrating Mass in San Salvador.

With the revolutions over, Catholic churches are now a mix of traditional and progressive members. Protestantism, which came to South America largely through immigration and conversions made by evangelical missionaries from the United States, is especially strong in Argentina, Chile, and Brazil. There has been a marked rise in Pentecostalism throughout the continent. A land where Catholicism dominated now sees Catholicism and Protestantism in competition. There are now more than 549 million Christians in Latin America.[36]

On March 13th, 2013, the first pope from the Americas was elected. He is Jorge Mario Bergoglio, a Jesuit from Argentina, now known as Pope Francis. He is a humble man of the poor, and his election brings much hope for much-needed reform of the church.

Asia – Christianity has had a presence in Asia for many centuries. The stories in these countries have often been of persecution, especially under Communism in China and the imperial rule during the seventeenth and twentieth centuries in Japan. The following is an overview.

- *China*

Christianity first came to China in the seventh century. In the sixteenth century Matteo Ricci (1552–1610), a Jesuit Renaissance man, mathematician, linguist, mapmaker, musician, theologian, and lawyer, traveled to China. Ricci honored the teaching of Confucius and Chinese religions, and demonstrated their connection with Christianity. By 1640 there were about 70,000 Christians in China.[37] Eventually, more conservative priests came on the scene.

They persuaded Rome that Ricci and the Jesuits had been too accommodating to pagan China. The emperor retaliated by outlawing Christianity, and the Western missionaries and Chinese priests were deported. Still, the Christian communities struggled on under lay leadership. Amazingly, the Christian communities survived and by the nineteenth century, when the colonials arrived, they numbered more than 200,000.[38]

The nineteenth century saw a strong interest in foreign missions in Europe, and many sought to convert the Chinese. They were met with suspicion that they would undermine the social order. Christians gained influence by opening schools, hospitals, and shelters, but during the anti-Western Boxer Rebellion in 1900 many missionaries and thousands of Christians were killed. After the tensions calmed, more missionaries arrived. The Protestants flourished and more Chinese Catholic lay people advanced their church's cause. After surviving several anti-imperialist persecutions, the churches were attacked when the Japanese occupied China (1937–1945). Christians supplied much relief for refugees during that time.

After WWII, Christians in China expected to flourish, but the victory of Communism in 1949 shattered their hopes. Foreign missionaries were expelled, and the government took control of the churches. During several national campaigns from 1958–1969, all religious activities were forbidden, Christian homes were ransacked, religious books were burned, and religious leaders were publicly humiliated and forced into labor camps.

Since the Mao Tse-tung period has ended, there has been more religious freedom in China. Most religions are allowed to function, but they have to register with the government and are carefully monitored. The Catholics maintain registered churches as well as underground churches, which have often been subject to oppression.[39]

• *Japan*

In Japan, the sixteenth-century Jesuits were able to befriend the rulers and, like Ricci in China, adopted the dress and lifestyle of the Japanese. The Franciscans and Dominicans preferred to work with the poor and did not approve of adaptation to the culture. Some of the poor resisted the government, and in 1637 more than 37,000 Christian peasants were killed in an uprising. Conflicts arose among the religious orders; ruling shoguns deemed this a danger to national security. Christianity was banned and subjected to long periods of persecution. "Picture stepping" was common, whereby images of Jesus and Mary were placed in the street, and citizens were required to trample on them.

During WWII foreign bishops had to resign and foreign missionaries were either imprisoned or expelled. Since the post-war period there has been some recovery, but Christianity is still viewed by many as a religion foreign to the Japanese culture. It keeps a low profile and uses its schools and excellent universities to speak for it. Catholics, Protestants, and Orthodox Christians comprise about 2 percent of the Japanese populace.

> **See Hidden Christians:**
> http://www.youtube.com/watch?v=xkw7UeX_p74

Christianity in Asia is gaining strength today. The Catholic Asian bishops have become a strong pastoral and theological force in the church, and many Catholics worldwide look to them for leadership for the future. There is much vitality in the Christian churches in Japan, China, Korea, Vietnam, and the Philippines. Often, their method for spreading the gospel is through service of others, especially the poor.[40]

Asia, which includes five regions of the Orient and the Middle East, contains two-thirds of the world's population. Even though

Christianity began in the Middle East in Palestine and Syria, the Christian population in Asia is about 3 percent of the population, more than 352 million.[41]

MODERN MOVEMENTS IN CHRISTIANITY: CONNECTING CHRISTIANITY WITH WORLD ISSUES

Today many Christians sustain only a nominal relationship with their church or congregation. Church attendance has dropped significantly in Europe and the United States among both Protestants and Catholics. The major challenge for Christianity seems to be the growing division between those who are traditional and wary of change, and those who are progressive and want to adapt more vigorously to the changing times. There is a marked growth in the so-called mega-churches, as well as in Pentecostal and evangelical communities, in the United States.

The Black churches in the United States are showing new vigor and, while still being concerned about equality for African Americans, have also turned their attention to problems within their neighborhoods, such as drugs, gang violence, struggling single mothers, and the vulnerability of Black families.

We have seen that Christianity is a dynamic, evolving religion that often had to adapt to different cultures and address many issues at a time. Worldwide, there are many Christians who have linked their religious beliefs with global needs. Internationally, churches and congregations serve refugees and poor communities in developing countries with food, shelter, clothing, and health care and are actively engaged in effecting societal change. In the United States, Christian organizations provide care for the elderly, soup kitchens and shelter for the homeless, day care for children, and food and clothing for the needy, as well as health care.

In the following section, we will examine Christian involvement in a few global concerns: ecology, peace, and the women's movement. We will single out some of the Christian values that are being applied to these issues and highlight individuals who are active in these arenas.

Christian Values and Ecology – Many date the beginning of the modern environmental movement with the publishing of Rachel Carson's book *Silent Spring* in 1962. This book was a study of the poisoning of the land and its life systems with chemicals, especially the much-used DDT. A standard bearer for Christians in this movement was Lynn Townsend White Jr., a cultural historian at the University of California at Los Angeles, who in 1967 charged that the Christian belief in "human's dominion over creation" is largely responsible for much of the degradation of the earth by industrialization, chemical manufacturing, and technology. White's charge brought an outburst of response from Christian thinkers, particularly among Protestants. In 1991 the Protestant organization World Council of Churches published a significant document on religion and ecology that was followed by powerful statements from American Baptists, Presbyterians, and Lutherans.

The Catholic Church's response to ecological concerns has been gradual. Vatican II did not address ecology, and even the official *Catechism of the Catholic Church* (1994) comments only obliquely on environmental dilemmas. Many national Bishops' Conferences, however, published valuable documents on local ecological concerns in the 1980s and 1990s. Pope John Paul II issued a strong statement on the "moral crisis" of environmental degradation in 1990. Pope Benedict XVI has written on ecology; some refer to him as the "green Pope." The Jesuits have made ecology one of their international priorities, and there is now extensive Catholic writing on ecology by both women and men authors and activists. Newly elected Pope Francis has taken his name from St. Francis, the "patron saint of ecology," and will no doubt be concerned about care for the environment.

The following are some Christian values that are today being linked to ecology.

• The Goodness of Creation

Genesis 1 contains a liturgical hymn with the refrain "God saw that it was good." Biblical scholars tell us that this is a goodness of usefulness, a goodness that serves God's plan and purpose. At the same time, all things are linked together and work together to achieve God's purposes. In this perspective, all things are good because they belong to God. "The earth is the Lord's" (Psalm 24:1). All are good because God is in all: "The spirit of the Lord has filled the world" (Wisdom 1:7).

• Humans as Co-creators

As mentioned earlier, Jesus was a Jewish reformer, and from the many images of God in the Hebrew Scriptures, he restored the Abba image—the creative, loving, healing, and forgiving father God. Rather than the warrior, punishing God, Jesus revealed the nurturing, freeing God—the God whose kingdom or reign is one of peace and love.

The mission Jesus gave his followers was to continue spreading the good news of this kingdom, which includes the health of the earth and all its creatures. Thomas Berry puts it bluntly: "If there is to be any acceptable human future, the grandeur of the planet must continue to flourish."[42]

• Concern for the Poor

Christianity follows its founder, Jesus, in his concern for the poor of the world. It is clear that the poor, especially deprived women and children, suffer the most from distressed environments. The underprivileged usually live closest to dumping grounds, breathe polluted air, and drink contaminated water. As we have seen in the horrible experience of Hurricane Katrina, Haiti's earthquake, and the floods in Pakistan, it is the poor who are most at risk. Christians are becoming more aware that service of the poor includes not only aiding survivors of natural catastrophes, but also becoming proactive in changing laws. Global warming is now on the agenda of many Christian environmentalists.

Christians in Action

Dorothy Stang (1931–2005) – Born and raised in Dayton, Ohio, "Dot" left for the convent after high school, wanting to serve in the missions. Eventually she was sent to Brazil, where she ministered for thirty years, defending the poor who were oppressed and displaced from their land by the loggers and cattlemen. She was a well-informed and formidable opponent of those who wanted to destroy the rainforests in her area of Brazil. She wore a T-shirt that said: "The death of the forest is the end of our life." Even though she knew that she was on a hit list, she remained at the side of her people until she was gunned down by a hired killer.

>> *Visit the Dorothy Stang Memorial and listen to Sam Clements discuss his meetings with Sister Dorothy to learn more about her position on her faith and the environment of the Amazon:* http://www.youtube.com/watch?v=Z-PK7uWfnBA *and watch his film*

They Killed Sister Dorothy: http://www.youtube.com/watch?v=sVaqqPURp1U

Prince Philip (b. 1921) –Prince Phillip is the husband of Queen Elizabeth of England. Philip served in the British Navy in WWII and is an accomplished horseman and yachtsman. He founded and has been dedicated to the **Alliance of Religion and Conservation**, and hosts its conferences in Windsor Castle. The ARC's aim is to assist the major world religions to develop environmental programs based on their core teachings, beliefs, and practices. The organization is now joined with the United Nations to develop faith-based plans to serve ecology.

Christian Values and Peace – Christian theology teaches that Jesus opposed violence, telling his followers to turn the other cheek and love their enemies. In his Sermon on the Mount, he proclaimed, "Blessed are the peacemakers," and he insisted that his disciples put away their swords when they tried to prevent his arrest. Scripture scholars point out that Jesus was not passive, however, but courageously confronted evil.

The history of Christianity has been marked with much violence, especially during the Crusades, the Inquisition, the conquests of the New World, the religious wars between Protestants and Catholics, the American Civil War, and two World Wars. Nevertheless, in the modern period there has been a profound interest in peace and nonviolence.

Dorothy Day, the founder of the Catholic Worker Movement, which serves the homeless and laborers, lost many of her closest followers when she protested America's entrance into WWII. Cesar Chavez, who spent most of his life fighting for justice for migrant farmworkers, insisted on using only peaceful means in marches and boycotts. Martin Luther King Jr., a Baptist pastor who was influenced by Gandhi's pacifism, used nonviolent tactics to liberate American Blacks from the Jim Crow laws and gain the civil rights due them by the American Constitution. Both King and the Trappist monk Thomas Merton were serious critics of the Vietnam War and nuclear arms. The Jesuit Daniel Berrigan was imprisoned for his destruction of draft records during the Vietnam War and continues to protest wars today. William Sloane Coffin Jr. (1924–2006) was a Presbyterian leader and a highly visible peace activist, who led protests against the wars in Vietnam and Iraq. He also demonstrated nonviolently to gain rights for blacks and homosexuals. Archbishop Desmond Tutu peacefully challenged apartheid in South Africa and was instrumental in its removal. Pope John Paul II was committed to nonviolence and, just before his death, objected to the invasion of Iraq. Helen Prejean is a Catholic nun who has effectively lobbied against capital punishment in the United States. These are only a few of the many Christians who have protested the violence of this era and who have advocated nonviolent means to resist oppression.

>> **See Cesar Chavez:** *http://www.youtube.com/watch?v=rj4ya_Gyq80*

**Watch Dr. King's final speech and note the number
of biblical references he uses to make his points:**
http://www.youtube.com/watch?v=smEqnnklfYs

The following are some of the Christian values that have inspired advocates for peace and nonviolence:

• *The Sacredness of Human Life*

The Christian tradition holds that humans are created by God, in the very image of God, and that human life is to be held sacred. Humans are called to share in God's own life (grace), in this life on earth and in life hereafter. Therefore, killing is forbidden in both the Hebrew and Christian Scriptures.

Christians believe that humans have the right to be free from oppression and violence. Humans have the right to food, drink, shelter, health care, education, and a living wage. From this perspective, peace is not simply the absence of war, but a condition where human dignity and rights are recognized and preserved. Peace is therefore concerned with the just rights of all human persons.

• *Nonviolence*

In the Christian tradition there is a range of attitudes toward nonviolence. The just war theory has been prominent in Christianity since the fifth century. War is allowed on the conditions that decisions are made by a competent authority, that there is a just cause, a right intention, a reasonable chance of success, and a proportionate use of violence, and that war is carried out only as a last resort.

Many scholars today question whether the just war theory can be useful in the event of a nuclear war. Many Christians also believe in nonviolence, ranging from the absolute non-use of violence in any situation to those who allow for legitimate self-defense. In the case of activists such as Gandhi, who was influenced by the Christian gospels, and King, nonviolent resistance was a planned strategy that required vigorous training and was aimed at confronting unjust systems and advocating for freedom.

Christians in Action

Archbishop Desmond Tutu (b. 1931–) – Desmond Tutu is an Anglican archbishop in South Africa. During the 1980s, he heroically but nonviolently opposed apartheid, the governmental system that separated blacks and whites and severely oppressed blacks. Many black leaders, including Nelson Mandela, were killed or imprisoned during the resistance. Tutu was in the front lines in the protest and in 1984 was awarded the Nobel Peace Prize for his efforts in liberating his people.

>> *Listen to Archbishop Desmond Tutu on leadership:*
http://www.youtube.com/watch?v=lrCeVwwuOXc

Peggy (1942–) and Art Gish (1939–2010) – Peggy and Art Gish are Mennonites who have spent their lives as civil rights and peace activists and members of **Christian Peacemaker Teams**. Both have been popular speakers on campuses. Peggy was in Iraq before and during the invasion, advocating for the Iraqi people. She survived a kidnapping there. Peggy has written *Iraq: A Journey of Hope and Peace*. Art worked extensively for peace in Israel, especially in Hebron. He wrote *Hebron Journal: Stories of Nonviolent Peacemaking* and proclaimed: "We're supposed to be glad and rejoice because so many of our enemies were killed. As a Christian, I find that repulsive...Over one million Iraqi people have died in response to 3,000 being killed on 9/11. Isn't there a better way? Yes. The way of Jesus."

Ironically, after standing in the way of tanks and bulldozers, Art was killed in 2010 in a tractor accident on his organic farm in Ohio.

>> **Meet Peggy and Art Gish and learn more about their work for civil rights and peace:** http://www.youtube.com/watch?v=8Eezf5AsWrk&NR=1

Bishop Thomas Gumbleton (b. 1930–) – Thomas Gumbleton is a former auxiliary bishop of the Detroit Roman Catholic Archdiocese who has been a leading voice for peace, justice, and civil rights in the United States. He co-authored the 1983 U.S. Catholic Bishops' Conference Pastoral Letter, "The Challenge of Peace," and was one of the first bishops to speak out against the Vietnam War. He is a founding member and past president of Pax Christi USA, the American Catholic peace movement. He is also a founder and former president of Bread for the World. Since becoming a bishop in 1968, he has traveled worldwide, protesting war and nuclear weapons. He has met with victims of war in Iraq, Afghanistan, Vietnam, El Salvador, Nicaragua, Guatemala, Israel, Palestine, Colombia, Haiti, and Peru.

Bishop Gumbleton has been an advocate for the full participation and the rights of women and homosexuals in the Catholic Church and has been a staunch advocate for victims of sexual abuse by the clergy.

>> **See Let's Drop the Bomb: Gumbleton on Nuclear Disarmament:** http://vimeo.com/11559690

Christian Values and the Women's Movement – As a Jewish reformer, Jesus called for changes in the divorce laws that had, until then, favored men, and he opposed the stoning of women who were unfaithful. He ignored many taboos by talking to and healing women in public, visiting them and instructing them in Torah in public and in their homes, dining with them, and even selecting them as disciples. These actions were unheard of in his day. Women were clearly present at his crucifixion and were the first witnesses to his Resurrection. Mary Magdalene is featured in the gospels as a key disciple, who actually funded Jesus' mission (Luke 8:1–3). In all four gospels, she is the first to witness to the Resurrection.

The gospels show great reverence for Mary, Jesus' mother. She is featured in the birth stories, at Jesus' first miracle at Cana, and at the foot of the cross. Paul, the apostle to the Gentiles, continued with this inclusive approach to women in the Christian communities. Although he stuck to some cultural practices, such as women covering their heads and not speaking out in church, a number of women were his closest associates in ministry. Priscilla worked side by side with him in Rome, Corinth, and Ephesus. Paul offers high praise for her and her work in leading the churches. Lydia helped Paul establish the church in Philippi. Paul calls Junia an apostle. At the end of his letter to the Romans, Paul cites ten women who are leaders of house churches.[43]

Jesus' practice of including women in ministry seems to prevail in the early church as virgins, widows, and deaconesses join the ministry of prophets, teachers, and apostles. As Elisabeth Schüssler Fiorenza, a leading woman biblical scholar, points out: "the Pauline literature and Acts still allow us to recognize that women were among the most prominent missionaries and leaders in the early Christian movement."[44] As we have seen, this radical Christian stand on the equality of women eroded in the face of Greek and Roman influences, especially after Christianity became integrated into the Roman Empire in the fourth century. Under the influence of Augustine, Jerome, and other church fathers, women were characterized as inferior, weak (inclined toward heresy, and therefore, no longer allowed to participate in ministry), and "dangerous and tempting" for the growing number of men committed to celibacy in religious life.

There were some exceptions to the exclusion of women in Christian history: Abbess Hildegard and some of the other powerful abbesses in Europe; Catherine of Siena, who was advisor to popes; and Teresa of Avila, a mystic and reformer of convents. However, many women mystics and healers were declared to be witches who performed orgies with the devil, and they were imprisoned or burned at the stake. (More than 100,000 women suffered such an execution.) The widespread devotion to Mary the mother of Jesus improved the feminine image, but this was often so cosmic and majestic that it was out of reach for the average Christian.[45]

Neither Reformation had a significant effect on the role of women in the churches. Luther stressed fellowship among the sexes and was an advocate for the education of women, but did little with regard to their role in religious leadership. Calvin wrote positive theology about women, but on a practical level was patriarchal. In the seventeenth century there were radical Protestant Reform movements, such as the Quakers, and some women began teaching and leading services. They were often punished or even imprisoned by the authorities.

In the modern era the role of Christian women began to change. In 1890 Elizabeth Cady Stanton organized The Woman's Bible, which set out to eliminate sexism in the Scriptures. During this period some congregations began to ordain women, but that did not gain momentum until the mid-twentieth century. In 1899 Frances Willard wrote: "It is men who have given us the dead letter rather than the living gospel. The mother-heart of God will never be known to the world until translated into terms of speech by mother-hearted women."[46] Ever since, countless women scholars have offered enormously valuable feminine critiques and interpretations of the Scriptures and indeed of the entire Christian tradition.

Today there are many leading women theologians, biblical scholars, and religious leaders in both the Protestant and Catholic churches. Women write as Christian feminists, as "Womanists" representing the Black experience, and as "Mujeristas" representing the Latinas. Globally, women Christians have found their voice.

Ordination to ministry is well-established in nearly all the Protestant churches, as well as ordination to the level of bishop. In 1994 the Anglicans accepted women's ordination. Many Anglicans and Episcopalians object to women's ordination to priesthood, and most especially to the recent ordination of openly gay men and women as bishops.

Eyebrows were raised in the Catholic Church when it was discovered that during the Soviet occupation of Czechoslovakia, several women were ordained Catholic priests in the underground church. One of the ordained women, Ludmila Javorova, has publicly described her ordination and given addresses on her ministry. The Pontifical Biblical Commission in Rome in 1976 declared that the Scripture does not give sufficient evidence to exclude the possibility of the ordination of women.[47] The Vatican remains adamant that women may not be ordained, and Pope John Paul II declared definitively on the matter. The main argument is that Jesus chose only men for his apostles and that these male-only bishops are the successors of the apostles who carry forth the priesthood. In this argument, Jesus' own will precludes the ordination of women, and this determination cannot be changed by the church. Movements such as Women-Church, the Women's Ordination Conference, and Roman Catholic Womenpriests, wherein more than one hundred women have been ordained so far by male and female bishops, are challenging this teaching and are willing to be subject to excommunication in order to prevail.

The following are some of the Christian values emphasized in the struggle for women's equality in the Christian churches and congregations:

• The Discipleship of Women

Scripture studies have clearly revealed that Jesus held great regard for the dignity of women and chose them to be his followers and to continue his mission. This same tradition was carried on in the early church until its gradual dismissal by a patriarchal culture and hierarchical structures within the church.

Reform movements have traditionally attempted to return to the earliest and purest practices in the Christian tradition. Modern biblical and theological studies indicate that such reform would mean offering women opportunities on all levels of teaching and ministry.

• The Equality of Women

The Scriptures, along with their patriarchal views derived from culture and male authors, also offer an inspired vision of women as prophets, leaders, mothers, and teachers. Feminine images are used for God in both the Hebrew and Christian Scriptures, and Wisdom is portrayed as a lady. Many religious leaders, both Protestant and Catholic, affirm the equality of women.

Modern culture also is rapidly recognizing the equality of women on all levels of society from family to military and corporate life. Suppression and violence toward women is recognized as unacceptable and even criminal. Many believe that if the churches are to play a prophetic and formative role in today's world, women are needed to be role models of leadership and ministry.

Christians in Action

Joan Chittister (b. 1936) – Joan Chittister is a Benedictine nun from Erie, Pennsylvania. She holds an M.A. from the University of Notre Dame and a Ph.D. from Penn State University. Sr. Joan is a popular international lecturer and a prolific writer on spirituality, peace and justice, and women's issues. She is a force in the Catholic Church for women's rights in ministry and social justice in the world. Joan is co-chair of the **Global Peace Initiative of Women**, an organization that has partnered with the U.N. to assist women in becoming peace builders.

>> *Consider Sr. Joan's observations on gender justice from her talk at the International Gender Justice Dialogue* http://vimeo.com/11123980 *and her presentation on the importance of interreligious forums at the Council for a Parliament of the World's Religions.* http://www.youtube.com/watch?v=CcxKZdJOxB4

Bishop Barbara Harris (b. 1930) – Barbara Harris grew up in Philadelphia. She attended business school and was hired by a firm where she eventually became president. Later, she studied for the ministry at Villanova University and continued her studies in England and Pennsylvania. She was ordained in 1980 and regularly consults with corporations on ethical questions. In 1989 Barbara was ordained as the first female Episcopal bishop in the worldwide Anglican Communion. She has been dedicated to prison ministry and questions of social justice.

》 *Consider the points in Bishop Barbara Harris'*
sermon on the role of Mary in the Christian faith:
http://www.youtube.com/watch?v=0tO0H9wxNPA

Bishop Patricia Fresen (b. 1940) – Patricia Fresen was born into a strict Catholic home, one of twelve children. After high school, she joined the Dominican Sisters and taught in their schools for seventeen years. At age 39 she studied theology in Rome and wrote extensively on liberation theology and apartheid. In South Africa she taught and became more aware of the evils of racism and sexism. She then studied in the United States and Canada and finally gained her doctorate in South Africa, where she taught in a seminary for seven years. In 1998 she took a post at a Catholic university in Johannesburg, where she became interested in women's ordination. Patricia was ordained a Catholic priest in 2003 in Spain and was required to withdraw from the convent and her teaching position. She was later ordained a bishop and now lives in Germany where she serves as a leader in the controversial movement called Roman Catholic Womenpriests.

》 *View Roman Catholic Womenpriests Bishop Patricia Fresen's*
homily: http://www.youtube.com/watch?v=UJQFDNMnAO8&feature=related.
What justification do you find within the Christian religion for the
women's ordination initiative?

SUMMARY

Christianity began with a small group of Jewish disciples who followed Jesus, a Galilean carpenter turned preacher and healer: a reformer who called his fellow Jews to return to justice, love, and kindness. After experiencing Jesus as being raised from the dead, these disciples came to recognize Jesus as the Christ, the anointed one of God and their savior. They were transformed from frightened and confused common folk to enthusiastic missionaries of Jesus' good news—the gospel—that the kingdom of God was present in the world and would ultimately prevail and reign over all. They followed the Master in calling people to repentance and conversion.

The community spread throughout the Middle East into Greece and as far as Rome. Due mainly to the work of Paul the Apostle, the gospel spread to non-Jews, known as Gentiles, who often had to meet secretly in house churches to avoid persecution. The movement spread rapidly for several centuries, largely in urban areas. Amazingly, Constantine the Roman Emperor was himself converted. Christianity was accepted and would eventually become the official religion of the empire. With the establishment of Constantinople as the capital of the empire, divisions between East and West were set in place.

Throughout the next few centuries heresies were confronted and excluded; official Scriptures, the Old and New Testament, were approved; and monasticism was established. By the sixth century the church had become a powerful force in the new Holy Roman Empire. In the ninth century, Islam conquered the Eastern churches but the West continued to spread the gospel throughout Europe. In the twelfth century the Crusades struggled to reclaim Palestine. Successful at first, the Crusades ended in failure.

In the Middle Ages and Renaissance, the Western Church flourished, building magnificent cathedrals and universities, and making impressive advances in music, art, and scholarship. Christendom was at its peak, but there were signs of decline. The Bubonic Plague severely weakened the entire culture of Europe, including the church. The clergy was decimated and the papacy corrupted. The Inquisition, which imprisoned or burned at the stake so many suspected of being heretics, alienated and terrorized the public. Superstition, fear of death, and the selling of indulgences to gain salvation were commonplace.

Reformers appeared: first Luther, then Zwingli and Calvin, and Henry VIII of England. The churches divided between Catholic and Protestant as the reform spread throughout Europe. The Catholic Church began its own reform, which culminated in the Council of Trent (1545-1563), but it was too late: the divisions were now established and still remain today.

The Christian churches still maintain their sacramental system with Catholics, both East and West, opting for seven and Protestants accepting only two.

All churches honor the Bible as the inspired Word of God and feature the Word at their Sunday services, along with sermons, prayers, and songs. Most Christians celebrate Christmas as the birthday of Jesus and honor the Holy Week that leads up to the major feast of Easter, which celebrates the Resurrection of Jesus Christ.

Today, Christianity is a world religion, which exists in nearly every country. During the last twenty centuries it has spread throughout Europe, where it now seems to be in decline. Christianity exists in India and Asia, where it is a struggling but vigorous minority. It is still strong in North and South America, where it is gaining new vigor among the growing population of Latinos. In Africa, Christianity is achieving amazing growth and vitality.

Today, some of the challenges that face the modern world are in the areas of ecology, peace, and women's issues. There are clear signs that Christians are reexamining their beliefs and values to see how these might address the pressing problems of our time.

CHAPTER 5 VOCABULARY

apostle - *A select disciple of Jesus*

baptism - *The ritual of Christian initiation*

ecumenism - *Dialogue among Christian groups*

Eucharist - *The Lord's Supper*

Gospel - *An early account of Jesus' life and teachings*

incarnation - *The belief that the Son of God became human*

Lent - *Forty-day preparatory time before Easter*

Orthodox - *The Eastern churches*

Pope - *The bishop of Rome and leader of Roman Catholicism*

Sacraments - *Key rituals of Christianity*

sins - *Acts that are morally wrong*

Trinity - *Belief that there are three persons in God: Father, Son, and Holy Spirit*

APPLYING CHRISTIANITY TO WORLD ISSUES

1. What values does Christianity bring to the table for world peace?

2. Do you think women have made progress toward equality in the Christian churches? Explain.

3. Do you think "creation theology," the theology that teaches that God created the world and placed humans as caretakers, can be useful in moving the churches toward concern for ecology? If so, cite specific things the churches can do.

TEST YOUR LEARNING

1. From what you know of the gospels and Jesus' era, what kind of a person do you think he was?

2. What are some of the central elements of Jesus' teachings?

3. How did Christianity come to be accepted in the Roman Empire?

4. What do we know about the authorship and the time of composition of the gospels?

5. Explain several of the controversies in the earliest councils.

6. What was the purpose of the Crusades? How effective were the Crusades?

7. Some say that during the Reformation the church came down "like a house of cards." What were some of the main cards that fell?

8. List some ways that the Eastern Orthodox churches differ from Roman Catholicism.

9. What were the main reforms of the Second Vatican Council?

SUGGESTED READINGS

Alberigo, G. A Brief History of Vatican II. *Maryknoll, NY: Orbis Books, 2006.*

Baum, W., Winkler, D. The Church of the East. *New York: RoutledgeCurzon, 2003.*

Brown, P. The Rise of Western Christendom. *Oxford: Blackwell, 2003.*

Burkett, D. An Introduction to the New Testament and the Origins of Christianity. *Cambridge: Cambridge University Press, 2002.*

Chadwick, H. The Early Church. *London: Penguin, 1993.*

Chidester, David. Christianity. *New York: HarperSanFrancisco, 2000.*

Crossan, John Dominic. The Historical Jesus. *New York: HarperCollins, 1992.*

Davidson, I. The Birth of the Church. *Oxford: Monarch, 2005.*

Harmless, W. Desert Christians. *New York: Oxford University Press, 2004.*

Hastings, Adrian, ed. A World History of Christianity. *Grand Rapids, MI: Eerdmans Publishing Co., 1999.*

Hill, Jonathan. The History of Christianity. *Oxford: Lion Hudson, 2007.*

Jotischky, A. Crusading and the Crusader States. *Harlow: Pearson Longman, 2004.*

Kung, Hans. Christianity. *London: SMC, 1995.*

Lippy, C. and others. Christianity Comes to the Americas. *New York: Paragon, 1992.*

MacCulloch, D. The Reformation. *London: Penguin, 2004.*

Patzia, A. The Emergence of the Church. *Downers Grove, IL: InterVarsity Press, 2001.*

Woodhead, Linda. An Introduction to Christianity. *New York: Cambridge University Press, 2004.*

Wright, A. The Counter-Reformation. *Aldershot: Ashgate, 2005.*

NOTES

[1] Linda Woodhead, An Introduction to Christianity *(New York: Cambridge University Press, 2004), 21.*

[2] *John Dominic Crossan,* The Historical Jesus *(New York: HarperCollins, 1992), 268.*

[3] *David Chidester,* Christianity *(New York: HarperSanFrancisco, 2000), 43.*

[4] *Ibid., 91.*

[5] *Note that the Eastern part of the empire, also known as the Byzantine Empire, ended in 1453 with the fall of Constantinople to the Ottoman Turks.*

[6] *Augustine Casiday and Frederick W. Norris,* The Cambridge History of Christianity I *(New York: Cambridge University Press, 2007), 73-75.*

[7] *Vivian Greene,* A New History of Christianity *(New York: Continuum, 2000), 39ff.*

[8] *See Carmel McCaffrey and Leo Eaton,* In Search of Ancient Ireland *(Chicago: New Amsterdam Books, 2002).*

[9] *Adrian Hastings, ed.,* A World History of Christianity *(Grand Rapids: Wm. B. Eerdmans Publishing Co., 1999), 115ff.*

[10] *Metropolitan Hilarion Alfeyev,* Orthodox Christianity *(Yonkers, NY: St. Vladimir's Seminary Press, 2011), 107ff.*

[11] *Paul Mojzes, "Orthodoxy under Communism," in* Twentieth-Century Global Christianity, *ed. Mary Farrell Bednarowski, (Minneapolis, MN: Fortress Press, 2008), 141.*

[12] *Ibid., 220.*

[13] *Hastings, 323.*

[14] *Website of the Orthodox Church in America: http://www.oca.org/ QA.asp?ID=52&SID=3*

[15] *Alfeyev, 7.*

[16] *Tim Dowley, ed.,* Introduction to the History of Christianity *(Minneapolis: Fortress Press, 2006), 230ff.*

[17] *Yves Congar as quoted in Thomas F. O'Meara,* Thomas Aquinas: Theologian *(London: University of Notre Dame Press, 1997).*

[18] *Diarmaid MacCulloch,* The Reformation *(New York: Viking, 2004), 41.*

[19] *Chidester, 267.*

[20] *http://law2.umkc.edu/faculty/projects/ftrials/luther/lutheraccount.html*

21 *Ibid., 311ff.*

22 *MacCulloch, 40ff.*

23 *Ibid., 230-247.*

24 *Woodhead, 270-273, 289-290.*

25 *Kathryn Muller Lopez and others,* Christianity *(Macon, GA: Mercer University Press, 2010), 211.*

26 *Chidester, 46.*

27 *Gerard Mannion and Lewis Mudge, eds.* The Routledge Companion to the Christian Church *(New York: Routledge, 2008), 43.*

28 *Donald Senior,* The Passion of Jesus in the Gospel of Mark *(Collegeville, MN: Liturgical Press, 1991), 20ff.*

29 *Gerd Ludemann,* Paul: the Founder of Christianity *(Amherst, NY: Prometheus Books, 2002), 213ff.*

30 *Brennan Hill,* Exploring Catholic Theology *(Mystic, CT: Twenty-Third Publications, 1995), 282.*

31 *It was estimated that 15 million slaves were exported between 1500 and 1870. See Paul Lovejoy,* Transformations in Slavery *(New York: Cambridge University Press, 2012), 24ff.*

32 *Todd M. Johnson and Kenneth R. Ross,* Atlas of Global Christianity 1910-2010 *(Edinburgh: Edinburgh U. Press, 2009), 9.*

33 *Ibid., 335.*

34 *Hastings, 455.*

35 *Johnson and Ross, 9.*

36 *Ibid.*

37 *Jonathan Hill,* The History of Christianity *(Oxford: Lion Hudson, 2007), 304.*

38 *Hastings, 384.*

39 *See Jean-Paul Wiest, "Catholics in China," in Mary Farrell Bednarowski, ed.,* Twentieth-Century Global Christianity *(Minneapolis: Fortress Press, 2008), 31-44.*

[40] *See Peter C. Phan, "The Church in Asian Perspective," in Gerard Mannion and Lewis S. Mudge, eds.,* The Routledge Companion to the Christian Church *(New York: Routledge, 2008), 275-291.*

[41] *Ibid.*

[42] *Ibid., 209.*

[43] *See Galatians 3:28 and Hans Kung,* Christianity *(New York: Continuum, 1995), 121ff.*

[44] *Elisabeth Schüssler Fiorenza,* In Memory of Her *(New York: Crossroad, 1984), 186.*

[45] *Kung, 453.*

[46] *Allyson Jule and Bettina Tate Pederson, eds.,* Being Feminist, Being Christian *(New York: Palgrave Macmillan, 2006), 139.*

[47] *Pontifical Biblical Commission, "Can Women Be Priests?"* Origins 6:6, NC News Service, 1976, 92-96.

Islam

Mohamed Abdul-Azeez is an **imam** (leader) of an Islamic community in Sacramento, California. He sometimes finds being a Muslim in the United States a bitter experience. He hears of friends being laid off from their jobs because of their religion, learns about Muslims being called *ragheads,* or reads of someone tearing the scarf from a Muslim woman's head. One time an anonymous man called him on the phone, ranted and raved about Muslim terrorists, asked him, *"Why are you killing our sons and daughters in Iraq and Afghanistan?"* and then hung up. Every time a terrorist incident occurs, such as the 2009 shooting at Fort Hood army base near Killeen, Texas, Mohamed warns his wife to stay home that day, keeps his children out of school, and tells his congregation to be careful. Mohamed says that since the al-Qaeda attacks on September 11, 2001, many Americans have stereotyped Muslims as Arab terrorists who want to kill innocent people. (An **Arab** is any member of the Arabic-speaking peoples of the Middle East and North Africa. While most Arabs are Muslim, some are Christian.) He says that few realize that many Muslims are not even Arabs. Many are Asian (Indonesia has the largest Muslim population in the world), Russian, Chinese, African American, and even Native American. He points out that few know that one-third of the slaves that were brought to the United States from Africa were Muslims. The imam says that many Muslims are our neighbors, peace-loving people, who work as lawyers, doctors, nurses, teachers, cab drivers, and homemakers. They love their families and this country.

Mohamed points out that his people in the United States are engaged in two battles: one within Islam to overcome those committed to violence; and the other to alleviate the suspicion and even hatred that many of their neighbors have toward Muslims. His mission is to help overcome the ignorance that many have about Islam and win the hearts and minds of his fellow Americans to understand and respect Muslims, most of whom are dedicated to peace and service of others.

More than 1 billion people in the world are Muslim, one-sixth of humanity. Muslims live in fifty-six different countries and are part of cultures in India, China, Europe, and the United States. Only twenty percent of the world population of Muslims are Arabs. Islam is the second-largest religion in Europe and is growing rapidly. Soon Islam will be the second largest religion in the United States. Yet many in the West know very little about Islam. To discover **Islam**—a word meaning submission to the word of God—in its true light we must explore the background and life of its founder, Muhammad; study the revelations compiled in the **Koran** (or **Qur'an**) and the spiritual practices known as the **Five Pillars of Islam**; investigate the divisions of Islam, its worldwide growth, the factions and challenges that Muslims face today; and explore the inspiring work Muslims are doing in the areas of ecology, peace, and women's issues.

THE PROPHET MUHAMMAD

Muhammad was born around 570 CE in Mecca, an arid basin city in the Arabian Desert. Though remote, Mecca prospered for two reasons. First, it was a safe and refreshing refuge for caravans carrying their wares between Syria in the north and Yemen to the south. Secondly, Mecca had the **Kaaba**, a revered, cube-shaped building considered the most sacred site and a place of pilgrimage by Arabs, where the images and symbols of many gods and goddesses of Syria, Egypt, Africa, and even images of Jesus and Mary, could be worshiped.

Mecca had become a thriving oasis, where merchants and camel drivers could wash away the dust and sweat and rest and refresh themselves and their camels. It was a place to put down their weapons, enjoy old friends, compare goods, and do some haggling and trading of the spices, silks, linens, and other exotic goods they had brought on their camels. Mecca was also considered to be a place of pilgrimage, where the nomadic Bedouins and merchants on caravan could be inspired by singing and dancing as they circled the Kaaba, worshiping and praying to their tribal gods or goddesses.

Tradition says that the Kaaba had been built by Adam, rebuilt by Noah after the flood, and rediscovered by Abraham when he visited his wife Hagar, whom he had banished along with their son, Ishmael, answering a test from God. The Arabs believed that it was Ishmael, not Isaac, whom Abraham was about to sacrifice at Yahweh's request, when Yahweh intervened and saved the son. Ishmael would live on and become the father of the great nation of Arabs. The Kaaba is made of local stone and is cube-shaped, with its four corners pointing approximately in the cardinal directions, indicating that this is the center of the earth. The outside is covered by black and gold cloth. The interior is marble engraved with passages of the Koran and decorated with green and gold perfumed cloth. The Kaabah contains a large, black stone, which is thought to be a meteorite sent to unite heaven and earth.

Mecca was controlled by the Quraysh tribe, which prospered by controlling the oasis and pilgrimage center. Though Muhammad was a member of the Quraysh tribe, he did not benefit from its success and prosperity because he was born into a poor family clan, the Hashim, and was orphaned at an early age.

The Life of Muhammad –

» **View Biography of Prophet Muhammad (Saw) 1 of 6 at**
http://www.youtube.com/watch?v=BZCbToxL6Bs
and 2 of 6 at *http://www.youtube.com/watch?v=klVixCAGcyk &feature=related*
for additional background.

The Koran says little about Muhammad. Most of what we know was composed by scholars several hundreds of years after the Prophet's death.[1] Tradition tells us that Muhammad had a difficult early life. Muhammad lost
his father before he was born, and his mother died when he was six. He was
raised first by his grandfather and then by his uncle Abu Talib, a merchant.
His orphan status later gave him empathy for the poor and outcast.

Muhammad grew up working for his uncle on caravans and became
known as a responsible and trustworthy employee. On one trip with his
uncle, the young Muhammad met a Christian monk named Bahira in
Syria. The monk predicted that the boy was destined for greatness and
would be a prophet of God.

When Muhammad was twenty-five, Khadijah, a widow fifteen years
his senior, hired him to work in her caravan business. The young man
apparently caught her eye, and once he proved himself trustworthy, she
sent a proposal of marriage. Muhammad and Khadijah enjoyed a happy
marriage and partnership and had six children: two sons who died in
infancy and four daughters. They also adopted a male slave, Zayd, and
another boy named Ali, the son of his uncle Abu Talib, who had helped
raise Muhammad and had then fallen on hard times.

Muhammad became troubled at the political and religious chaos he
saw around him. His caravan trips to the north revealed the endless divisions and brutal wars that were going on between the two empires that
had dominated the region for many centuries. One was the Byzantine
Empire that stretched east along the Black Sea into Iraq, south into Palestine, Egypt, and North Africa, and to the west into parts of eastern and
southern Europe. The other was the Persian Empire that took a large

swath of the Middle East and stretched east to India.

Mecca was also in turmoil. Pilgrimages and abundant trading had made many of the Quraysh tribe prosperous, greedy, and centered on pleasure. The poor, including many widows and orphans, had been driven to the outskirts of the city and the mountains, where they often lived in squalor. Blood feuds and revenge killings were rampant among the clans and tribes. Women had few rights, were traded as property, and often spent their lives serving men. Daughters were seen as liabilities and often were killed at birth or left outside in the elements to die.

Muhammad also experienced a spiritual hunger among his people. The scenes around the Kaaba were frenzied as Arabs worshiped the numerous gods and goddesses, but disappointment was in the air because the people did not seem to receive much help from their idols. And Allah, the creator, was a distant god who seemed to have little influence on their everyday affairs.

Muhammad seems to have been influenced by the rebellious Hanifs, a small Arab minority who insisted that Allah was the only god and who refused to participate in the idolatry at the Kaabah.[2] Many Muslims believe that this insight corresponded with his conviction that human nature innately was drawn to but one God.

Muhammad would have been familiar with local Jews, who were descendants of Babylonian exiles or perhaps driven from Jerusalem by the Romans. He also would have come across Christians, some descendants of missionaries, some hermits of the desert, and others cast out as "heretics" because of their unorthodox beliefs about Jesus and Mary. It could be from them that Muhammad heard about the Hebrew and Christian biblical stories. Like many Arabs, he would have been hopeful that, in the manner of Jews and Christians, his people would one day have their own prophets and inspired scriptures. Little did he know, in time, he would be chosen to be the Messenger and final prophet of Allah!

Muhammad was devoted to his own spiritual life. Each year he went to Mount Hira for a retreat in a cave. On these retreats he would meditate and at times experience deep dreams. On one of these retreats, when Muhammad was about forty years old, he felt overpowered by a spiritual force and was told:

Read[3] in the name of your Lord who created man from an embryo;[4]
Read for your Sustainer is the most bountiful.[5]

Muhammad heard about a benevolent God, who had created humans and wanted to be intimate with his people and instruct them. He heard of the arrogance and rebellion of his people, and learned that Allah sees all, wants to draw near, requires adoration, and warns that those who resist him will be dragged to hell.

This extraordinary experience threw Muhammad into confusion and near despair. Was he being deceived by a *jinni*, a supernatural spirit that Arabs believed could either trick or help them? Terrified, he fled from the cave to the top of the mountain, where he encountered the "spirit of revelation," which he learned was the Archangel Gabriel.[6] Muhammad wanted nothing to do with all this, fearing that he would be viewed as a charlatan. He became extremely disturbed by his experience.

Muhammad left the mountain and returned home, where he sought refuge in the arms of his wife, Khadijah, who wrapped him in a cloak, comforted him, and assured him that he had not gone mad as he feared. She tried to convince him that he was too good a man and that God would not trick him. His wife then sought advice from her cousin, Waraqa, a former Hanif now turned Christian. Waraqa told Khadijah to tell her husband to be of good heart for he was indeed a prophet. Still, Muhammad remained skeptical and was unsure what to do next, especially since God then remained silent for a long period of time.

Both Muhammad and his wife were becoming more doubtful, when suddenly he was grabbed again by a force and told:

I call to witness the pen and what they inscribe, you are not de-
mented by the grace of your Lord. There is surely reward unending
for you, For you are born of sublime nature. So you will see, and

*they will realize, who is distracted. Verily your Lord knows those
who have gone astray from the path, and He knows those who are
guided on the way.*[7]

Muhammad now knew that he had a special mission among his people. Muhammad's initial message was not concerned with the abuses at the Kaaba or polytheism, but with the true nature of Allah and a need for radical social reform. Muhammad was the Messenger for God, who was a good and loving Creator, "the most merciful," "the most generous," and not the remote, dispassionate Allah of the Arabs. This merciful and generous God was protective of the poor and underprivileged and, therefore, called the greedy and prosperous to turn from their selfish ways and to care for the needy, the orphaned, and the beggars. This, of course, came with a warning: the day of judgment was coming, and those who continued on their evil path would be cast into the fire.

At first, Muhammad shared his amazing message only with family and friends: Khadijah; his cousin Ali; the slave that he freed, Zayd; and his friend Abu Bakr, the wealthy merchant. All told, there were at first about forty "Companions" of Muhammad. They were called **Muslims**—those who submit to God. Even though Abu Bakr began to spread the revelation to his associates in town, there was little stir because the message of Muhammad was not yet threatening to the status quo in Mecca.

After three years, the Revelation began to evolve and now asserted: "There is no god but Allah, and Muhammad is God's messenger." Slowly the leading members of the Quraysh began to realize that Muhammad believed that he was speaking as the Messenger of God and challenging their beliefs and way of life. Muhammad was claiming to speak for God and was condemning their greed, tribal vengeance, and violence. Muhammad was empowering the poor and the outcasts, telling them God would provide for them. He was teaching about immortality in the afterlife, the last judgment, and condemnation to hell for evil deeds. Most important, he was condemning the worship of the idols at the Kaaba. Now

Muhammad had the attention of the local leaders of Mecca, who were gaining much wealth from the pilgrimages and ceremonies at the Kaaba. Muhammad had to be stopped!

The city leaders first tried to bribe Muhammad. When that didn't work, they ridiculed him. Muhammad was impervious to it all and insisted on standing near the Kaaba, preaching his message to the pilgrims. Finally, the Quraysh boycotted Muhammad, his followers, and his family from any social interaction and from buying or selling goods, even food and water. The clan was laid waste during the boycott. Then, just as the ban was lifted and things began to look better for Muhammad, his beloved wife, Khadijah, and his close relative and protector, Abu Talib, died. The Quraysh leaders knew now that Muhammad was vulnerable and that they could now move to eliminate this upstart.

Muhammad began to receive physical abuse in the streets and threats on his life. Still having his unique revelations, he felt isolated. At one point, he was somehow transported to Jerusalem, where he encountered Adam, Abraham, Moses, Jesus—all the great prophets of the past. This journey taught him that prophets are sent to all people who are brothers and sisters. Yet few in Mecca would listen to him. It was time to leave his hometown.

Fortunately, Muhammad had received an invitation to mediate disputes in Yathrib (the name would later be changed to Medina, which means "the City," short for the City of the Prophet), a group of villages 250 miles north of Mecca. Slowly, his followers began to filter out of Mecca, and then Muhammad followed, barely escaping an assassination plot and efforts of Bedouins to capture him and collect the high bounty on his head. This migration is known as the **Hijra** (the Emigration) and is still celebrated by Muslims.

Muhammad began to apply his revelations to the forming of a new community in Yathrib (Medina). He reformed the old tribal hierarchies, established a new equality, and called for forgiveness and reconciliation over the old blood feuds and retributions. As chieftain, he followed his revelations and abolished the exorbitant market taxes. He called for alms (**zakat**) for the poor, the homeless, and the orphans, as well as freedom for slaves. He also called for more equality for women in the areas of inheritance, dowries, and property rights. Muhammad built a small mosque where his followers could meet and pray. He linked up with the Jewish communities in the area, honored some of their feasts and prayers, and even asked his followers to face Jerusalem during prayer. This conciliatory mood did not last long, however, and Muhammad began to have serious disputes with some of the Jews. The prayers were now turned away from Jerusalem toward Mecca. He began taking some wives to himself and set up housekeeping in an area adjoining the mosques. All this time, and for the next twenty-three years, Muhammad would continue to have his revelations.

Knowing that he had to eventually face his enemies in Mecca, Muhammad began by attacking their caravans and capturing goods. In time, the Meccans sent an army to attack Yathrib (Medina). Though outnumbered, Muhammad defeated the Quraysh at Badr in 624 CE and drove them back to Mecca in disarray.

The next year, the Meccans attacked again. This time, they injured and defeated Muhammad at Uhud. Then, in 627 CE, during the Battle of the Trench, the Quraysh sent ten thousand troops to attack Medina. The Muslims, in preparation, cleared the crops and dug a huge trench around the city. They were able to force the Meccans to withdraw in defeat. A controversial part of this battle was Muhammad's order to execute seven hundred men from the Jewish Qurayza tribe, who had been charged with treason.

On the homefront in Medina, Muhammad began to lose support. His constant antagonism of Mecca and numerous battles put a tremendous burden on the place where Muhammad had ironically been brought to be a peacemaker. Moreover, he was aging, and enemies were increasing their attacks.

Muhammad was gaining a reputation as a powerful military leader. He continued his raids on Jewish settlements and on Meccan caravans. The Prophet was becoming a menace to the Quraysh, but gained much support from the neighboring Bedouins. Eventually, Muhammad was able to lead a pilgrimage into Mecca, forcing the helpless Quraysh to stand aside without firing an arrow. Once at the Kaaba, he destroyed all the statues and images of the gods and goddesses because of his belief that there was but one God, Allah. He left only the images of Jesus and Mary. He solemnly proclaimed that the Kaaba was the house of the one God, Allah.

The Death of the Prophet – Two years after his triumphant cleansing of the Kaaba in Mecca and his establishment of the Muslim tribes as one community in a new political and religious order (**ummah**), Muhammad returned to Medina. He attempted to resume his peaceful life near the mosque, but apparently was in poor health. One day in 632 CE, Muhammad suddenly withdrew from the prayer service, returned to his room, collapsed, and died. Some would not believe that the great Prophet could die, but it soon became evident that he was only a mortal and had truly passed away. In the Koran he says:

I am a human being like you. It is revealed to me that your God is one God.[8]

He had named no successor and now the challenge would be to preserve Muhammad's message and his mission.

THE KORAN AND ISLAMIC TEACHINGS

Muhammad is believed to have been nearly illiterate, so when he received his revelations, he memorized them and then recited them to his followers, who either committed the revelations to memory or wrote them down. Eventually there was an explosion of interest in the words of the Messenger. The Arabs loved language and being a largely oral culture prided themselves on their Arabic poetry and stories. With regard to Muhammad's revelations, his followers believed that they were receiving not only beautiful words, but the very words of God.

After the Prophet's death, the first **caliph** (successor or deputy), Abu Bakr, decreed that all oral memorizations of the Koran and all the written versions be gathered into one text. The third caliph, Uthman, seeing that various versions were still around, ordered that a definitive version be compiled and the others be destroyed.[9] The text that is used today is divided into 114 **surahs** or chapters.

The chapters are not chronological, but are arranged according to their length—the longest in the first section, the shortest in the last section.

>> *Listen to Beautiful Qur'an Recitation by Tawfeeg As Sayegh at:*
http://www.youtube.com/watch?v=sCvraPVZM9Af

Muslims believe that the Koran is the "Speech of God," miraculously revealed to the Prophet Muhammad. They believe that to invoke the Koran is to invoke God. To be authentic it must be in Arabic, and it can be either read or recited. Recitation is of great importance, and great honor is given to those who can memorize the Koran. (In Dubai, it is possible that some prisoners may be released early from a prison sentence for memorizing the Koran.) Children are taught to recite the text in school, and Muslims celebrate great occasions with recitations.

For the Muslims, the Koran is not a book, but the living word of God, which offers "a healing and a grace for the faithful."[10] The recited Word can give strength to the living, comfort to the dying, and solace to those left behind after a loss. Muslims experience in the haunting inner rhythms, undulating tones, and guttural sounds of the Arabic recitation a power that can touch their hearts deeply. For the Muslims, the Word watches over them, offers them guidance, and intercedes for them with Allah.

Muhammad respected the "people of the book," that is, Jews and Christians. Indeed, Islam was developed with "a creative interaction with other Near Eastern monotheistic faiths."[11] The Koran in fact contains stories from both the Hebrew and Christian traditions, though these stories often differ from the accounts in the Bible. Muslims believe that Muhammad's revelations stand as a purification of these traditions and the final corrective revelation beyond these two traditions. Muslims see the Bible as an account of God's revelations, written in different books by different authors. By way of contrast, Muslims believe that the Koran is the direct revelation as narrated by Muhammad. This is not to say that the Koran is not open to interpretation. There is a vast body of literature of interpretation of the Koran, and since there is not a central authority to give an absolute decision in these matters, there are many disputes over the correctness of interpretations.

In the later revelations received in Medina, Abraham is described as a Muslim, the founder of the Muslim community, and the person, who along with his son Ishmael, built the Kaaba. Moses is portrayed as liberator and lawgiver of his people, with many parallels being made between Muhammad and Moses. Significantly, the Lord tells Moses of the future coming of Muhammad the Prophet and calls upon the people to honor him and follow his light to attain their goal.[12] The Koran reflects a deep regard for Jesus, describing him as "a Spirit from God," born of a virgin, a teacher, prophet, and miracle worker, but not the Messiah.[13] The Koran denies that Jesus died on the cross, but acknowledges that Jesus will come on the Day of Judgment. The Koran clearly denies the divinity of Jesus and belief in the Trinity, since the Oneness of Allah allows for no other divine persons.

>> **At The Koran website browse the electronic translation of the Islamic living word of God to discover favorite verses:**
http://quod.lib.umich.edu/k/koran/browse.html

The Hadith – The **Hadith** is distinct from the direct revelations of Allah to Muhammad. It is a collection of the words and deeds of Muhammad, as well as those of his early Companions. They represent the Prophet's values, practices, and laws. There are thousands of Hadith. They range from "details of what is permitted and forbidden (legal Hadith), details of ritual action, theology, accounts of creation and eschatology, personal etiquette, descriptions of the Prophet's character and more."[14] Not all the Hadith are recognized as being authentic, even though great efforts were made early on to check the credentials of those who transmitted the Hadith. Hundreds of thousands of Hadith were carefully examined for their authenticity, and only several thousand were approved. The collection now exists in three sections: weak, good, and sound. It is generally accepted that many of the Hadith were shaped by later social and political situations. The Sunnis (eighty percent of all Muslims) accept six collections of Hadith. The Shia accept some of these, but also have compiled collections of their own.

>> *View* **Hadith: A Beautiful Hadith** *http://www.youtube.com/watch?v=5bXsYOFJDcE* *and* **Recitation of Hadith Beautiful** *http://www.youtube.com/watch?v=hTHH3NfOMpw* *to experience the poetry of the Hadith and learn more about the words and deeds of Muhammad.*

Shariah – Shariah refers to Islamic law as it developed from the Koran, the Hadith, later analogical arguments (**qiyas**), and from the consensus of legal scholars about particular legal situations. Actions are classified into five categories: obligatory, meritorious, indifferent, reprehensible, and forbidden.

Obviously, this process is open to much disagreement and can end in a diversity of views. In the Sunni world there are four main schools of legal thought, while the Shia has its own distinct schools.[15] The approaches range from the open and liberal to the rigid and legalistic. The **Ulamas** are scholars who are learned in Islamic matters of law and theology. They are the guardians of the tradition, and the ones who could express the consensus of the community. This role is carried out similarly by both Sunnis and Shiites. Imams in the Sunni tradition are religiously educated men in the community, who are engaged by the mosque to lead the prayers. In the Shia tradition, the imams are given higher stature and not only lead prayers, but are the rightful leaders of the community. These clerics sometimes guide in the background; at other times they rule with an iron hand. The resulting tensions will be discussed in detail later in the chapter when we address contemporary Islam.

THE FIVE PILLARS OF ISLAM

The **Five Pillars of Islam** are mentioned in the Koran and developed in the Hadith. They have been organized into a schema of five actions or rituals that define what it means to be a Muslim. By their very nature, the pillars reveal that Islam is primarily a religion of action, and then a religion of belief. They bind Muslims together as a global community that prays, fasts, and offers alms together all over the world.

The First Pillar: Profession of Faith (Shahada) – The prayer that professes the faith is the nucleus of Islam and is called the **shahada**. This prayer is taken from two sections of the Koran (47:19; 48:29). It is recited when a person converts to Islam, is said daily by Muslims, and is usually whispered in a baby's ear at birth. It professes: "I testify that there is no God but God. I testify that Muhammad is the Messenger of God." For the Muslim, nothing can be put in the place of the One God: not other gods, nor other vices such as pride or greed, nor other things such as money or additional material things.[16]

The Second Pillar: Ritual Prayer (Salat) – Muslims are obliged to pray five times a day: just before sunrise, noon, afternoon, sunset, and in the evening. One prepares by taking off the shoes and performing a ritual cleansing of hands, arms, face, and feet with water. While verses of the Koran are recited, the person goes through a series of movements that include standing, bowing, rising, and sitting. The prayer is said facing toward Mecca. **Salat** (prayer) can be said individually, but is meant to be a communal act in a mosque, where all pray before Allah in equality. When possible, the communal prayer is required for the noon prayer of Friday. Those who are ill, traveling, or unable to perform the prayers are excused.

Prayer is of the utmost importance to Islam. Prayer places one in the presence of Allah, offers due praise and thanksgiving to Allah, purifies the person of sin, and brings the grace and strength to be faithful.

>> **Watch the video Learn How to Pray (Salat)** *at* *http://www.metacafe.com/watch/2902594/learn_h_to_pray_salat/* **to hear the prayers and observe the actions of the supplicant.**

The Third Pillar: Giving Alms (Zakat) – The word zakat literally means "purification," so alms are given to purify one's possessions and to show mercy upon the poor. Muhammad strongly opposed greed and hoarding money, and he had a deep compassion for the poor. Personal possessions must be purified by paying one's dues to the community to support the less-fortunate. Muslims are obliged to give two and a half percent of their possessions to those in need. These alms are usually given to the mosque and then distributed.[17]

>> **View the video entitled: What is Ramadan?** **An Explanation by Yusuf Islam** *at* *http://www.youtube.com/watch?v=Ta8j1z5LkJo* **to learn more about meanings of this practice.**

The Fourth Pillar: The Month-Long Fast (Sawn) – This fast, which is believed to have originated during the revelation of the Koran, is held during the month of **Ramadan**, the ninth month of the Muslim year. This usually occurs in the late summer or early fall of the Western calendar. During this time the Muslim is obliged to fast from sunrise to sundown from food, drink, and sexual intercourse. Ramadan is a time for spiritual reflection and growth. The fasting is intended to give a heightened awareness of Allah's blessings and what those in need are experiencing. This is also a time for giving generously to the poor. It is an opportunity to be with family and friends for the breaking of the fast in the evening. The final night of Ramadan is an important feast, and a time for the community to enjoy a large celebration together.[18]

The Fifth Pillar: A Pilgrimage to Mecca (Hajj) –

>> *See a hajj and watch the story of an American Muslim's spiritual pilgrimage,* Inside Mecca: *http://www.youtube.com/watch?v=KFQHgdmJqjo*

Where possible, Muslims are expected to make a twelve-day pilgrimage (hajj) to Mecca once in their lifetime and to take part in the rites at the Kaaba and the surrounding areas. The official hajj takes place during the last month of the lunar year. More than a million pilgrims gather annually at this time in Muhammad's birthplace for the celebration. Certain preparation is needed before such a pilgrimage. Two seamless white garments have to be purchased for wear by men, as well as modest clothes for women who must also cover their heads. These garments symbolize purity and eliminate all social distinctions. Men shave their heads and trim their beards, and women clip several locks of hair. Debts are paid, affairs are put in order, and the complicated travel plans are made well in advance.

Before entering the Grand Mosque that surrounds the Kaaba, the pilgrims don their simple garments. As they enter the enclosure, a prayer is recited by the throng:

"Here I am, O God, here I am. Praise belongs to you, and blessing and power. You have no associate. Here I am."

The pilgrim is then swept up in a mass of people and moves around the Kaaba seven times, a traditional religious number that often stands for "fullness." The Kaaba, which was emptied of all idols by Muhammad, now stands as the House of God, the sanctuary of Allah, the center of the world. People from every corner of the globe join in this amazing and moving procession around the sacred dwelling place of Allah.

When the procession around the Kaaba is completed, the pilgrims proceed to the next ritual inside the mosque. This act consists of running back and forth between two structures built over two hills, Safa and Marwah. This running ritual symbolizes Hagar's desperate search for water between two hills after she was exiled with her son Ishmael.

Next, the pilgrims proceed about four miles to Mount Arafat, the site of Muhammad's last sermon. Here the pilgrims stand from noon until sundown reciting prayers, recalling the Prophet's last sermon, and standing before God as one would on the Last Day. After an overnight at this location, the pilgrims, recalling Abraham's rejection of the demon's temptations, stone three pillars that symbolize Satan, and then sacrifice sheep, cows, and lambs in remembrance of the call of Abraham to sacrifice his son before he is reprieved. The meat is given to the poor. On the twelfth day, the pilgrimage is finished. The pilgrims return to the Great Mosque and circle the Kaaba seven times again. The exhausting but exhilarating pilgrimage is complete. The pilgrims can now remove their pilgrim garments and set them aside until death, when they will serve as a shroud.[19]

ISLAM SPREADS

The death of Muhammad left the Muslim community without their Messenger from God and their leader. On the face of it, it seemed impossible to replace him or to plan what would be next. Desperately wanting to sustain the unity among the Arabs that Muhammad had brought about, the community chose Abu Bakr, an early convert and a Companion of the Prophet, to lead the community under the title of **caliph**. During Bakr's brief two year rule (632–634 CE), the Arab tribes who tried to withdraw from the covenant with Islam were conquered, along with the rest of Arabia. During that same period, the Muslims conquered the southern areas of Palestine and Iraq.

The next caliph, Umar (634–644 CE), conquered the rest of Iraq, Syria, Egypt, and western Iran. The Muslims also occupied Sicily during the ninth century until they were driven out by the Normans in the mid-eleventh century.

Asia Minor and Europe were stunned by this new, powerful, spirited movement that seemed to emerge from thin air. The leaders of the Byzantine Empire fled in disarray, and the Persian Empire collapsed in the face of this onslaught of Arabs who fought with such fervor. The pace of the constant conquest slowed, and by the eighth century the Islamic Empire included North Africa and Pakistan. In 711 CE the Muslims entered Spain and ruled from Cordoba and Granada for nearly eight centuries. They were driven out of Spain in 1492 by King Ferdinand and Queen Isabella. They also proceeded through Western Europe, but were stopped at the crucial Battle of Tours in 732 CE by Charles Martel, a Frankish military and political leader. In 1453 the Muslims conquered Constantinople and made it the capital of the Ottoman Empire, which ended after World War I. The city is now modern-day Istanbul.

The Bedouins, camel drivers, and merchants had come a long way. Just a century after the death of Muhammad, the sultans had become a royal elite, and their "local cult" had grown to be the dominant religion of an empire. The empire soon became larger than that of either Alexander the Great or Rome. Eventually, Islam would spread to the Balkans, western China, India, and Southeast Asia. As a result of such expansion, the Islamic traditions were deeply affected by the various cultures they encountered. There were now many different ways to be Muslim.[20]

The Muslims' motives for such conquest were to gain power, acquire more land, and seize enormous riches, either to keep for themselves or to share with their fellow Arabs in order to maintain their loyalty. Religious conversion does not seem to have been their prime motive. In fact, the Muslims discouraged conversion and were generally more tolerant of other religions than those whom they had conquered.[21] They did have some rules for these religions: They were not permitted to proselytize; they had to wear some identifying dress; and they had to pay a special tax.

This is not to say that asserting the prominence of Islam was not on their agenda. During the reign of Abd al-Malik (685–705 CE) the coin of the realm read:

*"There is no God but God alone without partner. Muhammad is
the Messenger of God who was sent with the religion of Truth to
proclaim it over other religions."*

》 **Watch the videos Empire of Faith: Parts 1 and 2 at**
http://www.youtube.com/watch?v=yX3UHNhQ1Zk
and http://www.youtube.com/watch?v=X1PxJomypQE
for greater depth on the spread of Islam.

This was also inscribed on the milestones along the newly built roads.[22]
And Muslims did, of course, welcome converts. Some Christians and
Jews saw Islam as a purification of their own religion; others saw that
the successful Muslims seemed to have God on their side. Perhaps some
converts simply wanted to share in the prosperity of the Muslims, and, of
course, escape the tax at the same time.[23]

The unity among the Arab tribes that Muhammad brought about did
not last after his death. The first source of division was the choice of his
successor. As mentioned earlier, Abu Bakr was appointed the first caliph.
This immediately upset those who wanted Ali, the Prophet's cousin and
son-in-law, to be appointed the leader. There was argument on how to
create the society of Islam. Cries of tyranny and apostasy were tossed
back and forth. The third caliph, Uthman (644–656 CE), was murdered.
Finally, Ali was chosen, only to be assassinated five years later. Party lines
were drawn.

The Shia – The **Shiites** became the party loyal to Ali. They argued that
the caliph should be a direct descendant of the Prophet, through Fatimah,
Muhammad's daughter, and his son-in-law Ali. **Shia** is the short form of
Shī' atu 'Alī, meaning the followers of Ali. Later, this tradition added that
the caliph should also be an **imam**, or divinely guided religious teacher.
They were pleased when Ali's son Hasan was appointed, only to find out
that he opted out to nurture his vast fortune and enjoy his thousands of

wives. (Hasan was rightly dubbed "the divorcer!") Meanwhile, Ali's other son, Husayn, struggled to gain the leadership, only to be beheaded at the battle of Karbala in 680 CE. Even today Husayn is revered as a martyr. His slaying is still graphically celebrated by the Shiites on the feast of Ashura, the day commemorating his death.

Husayn was succeeded by his son Ali Zayn al-Abidin. After this, the path of succession gets complicated, which becomes the basis for a number of sects among the Shia. The "Fivers" held that the fifth generation was the legitimate one for succession of a new leader; the "Seveners" held that the seventh generation was correct; and the "Twelvers" recognized the twelfth generation as legitimate for succession and looked for a "hidden imam," a new absolute leader to appear within that generation. This is the tradition that has prevailed in Iran in recent times, when in 1979 the Ayatollah Khomeini claimed to be the viceroy of this twelfth-generation person with the power to act on his behalf.[24] The Shia branch of Islam, which is considerably smaller than the Sunni branch, is considered to be more conservative. Shiites are a majority in Iran and Iraq. There are sizable populations in Yemen, Syria, Lebanon, East Africa, Pakistan, and Northern India.[25]

The Sunnis – The second and much larger group is known as the **Sunni**, or people of the tradition (**sunna**). The Sunnis do not hold that the leader of the community (**ummah**) must be a descendant of Muhammad, but can be any leader who follows and obliges the community to follow the Islamic laws and beliefs as taught by individuals with the reputation of being "learned men" (**ulama**). These scholars are expected to be experts on both the Koran and the Hadith.

Sunnis often note that they differ with the Shia more politically than they do religiously. At the same time, the Sunnis point to their Golden Age of intellectual and cultural development as a sign of favor from Allah. The Shia read this same history as a time when they were suppressed as a minority by oppressive and illegitimate rulers.

As noted earlier, Sunnis and Shiites have some differences when it comes to shariah or Islamic law. The Sunnis follow the Koran and the Sunna (accounts of what Muhammad said and did), as well as the techniques used by their jurists to decide how shariah should be applied in new situations. Finally, the Sunnis follow the consensus of their scholars (the ulama) who act as interpreters of the law. This is based on the Prophet's teaching that his community "will never agree on an error." The latter two sources are where the Shia differ. While they agree to follow the Koran and Sunna, the Shiites have their own traditions from Ali, as well as their own jurists and interpreters. So while the Sunnis and Shiites have often lived side by side in peace, they sometimes have disputes when interpreting the shariah.

The situation in Iraq has brought the Sunni-Shia contrast into focus in recent times. The United States and 37 other countries formed a coalition force that found itself caught in a conflict when its 2003 invasion of Iraq deposed Saddam Hussein, who was a Sunni. The Sunnis were a minority in Iraq, but they had dominated and often persecuted the more numerous Shiites. After the fall of Saddam, the Shiites seized the opportunity to gain power and retaliate against the Sunnis. Ever since, the coalition, which officially become United States Force-Iraq as of January 1, 2010, has been caught in the middle of the conflict between Shia and Sunni vying for power, as the disparate groups attempt to form a coalition government.[26]

The Shiites prevail in both Iran and Pakistan, but are a minority in many other countries. Today the Sunnis make up eighty percent of the world Muslim population and regard themselves as the mainstream and traditionalist branch. Sunnis make up the majority population in Saudi Arabia, Egypt, Turkey, Syria, North African countries, Pakistan, Afghanistan, Central Asian countries, and Indonesia.

The Kharijites — Another group of Muslims that developed in the late seventh century was the Kharijites, a rebellious faction that insisted rulers be sinless and just, as well as strictly follow the example of the Prophet

Muhammad. Rulers who do not measure up should be eliminated, and this resulted in the assassinations of some caliphs, including Uthman and Ali. The extremism and violence of the Kharijites have led some scholars to think they are the forerunners of Muslim extremists today.[27]

The Sufis – Sufism arose in the early stages of Islam among those who wanted to go beyond obedience to Allah and his commandments to experience a deeper spiritual love for the Deity that dwells within. The Koran speaks of the God within: "We know that his innermost self whispers within him: for We are closer to him than his neck-vein."[28] The Sufis maintained that they gained insights from the Koran and from imitating the mystical life of Muhammad. The story of Muhammad and his experience with an angel and his transport to Jerusalem (The Night Journey) were emphasized by these early mystics. Eventually, these Muslims were referred to as Sufi. (*Suf* means wool, and they wore woolen garments to symbolize purity.)

The Sufis practiced ascetic, spiritual, and liturgical exercises designed to deepen their devotion. They rejected the vanities of the world and were highly critical of the worldly ways of some Islamic leaders of their time. Sufi teachers warned against the attractive yet dangerous ways of the world. Some Sufis rejected marriage so that they could focus entirely on their relationship with God.[29]

The Sufi way is to love God for the sake of loving God, not out of a sense of duty or fear. Sufis follow the Islamic denial of original sin, doing penance for sin, or earning redemption through suffering. Their asceticism, rather, is done for the purpose of purifying the self in order to be united with God. Some of their ascetical practices may have been influenced by early Christian hermits. Some Sufi mystical beliefs may have been borrowed from the Eastern Christian churches. Most widely known are their Whirling Dervishes, graceful spinning dances used to bring on mystical experiences. Sufism still has a strong position in Islam, and its spirituality draws many converts in both Europe and the United States.

Sufis have been influential in spreading Islam to Africa, India, and the Far East. Rumi (1207–1273 CE), a great scholar and magnificent Sufi poet, is still well-known today and remains a hero for all Sufis. Sufis constantly remind Islam that it is a way of life, a way of the heart aimed at union with God.[30]

THE GOLDEN AGE

Gradually life in the vast Islamic Empire began to flourish. The East and West trade routes were controlled. Goods from Africa, Asia, India, and Northern Europe became available in the great capitals of Córdoba in Spain, Fez in Morocco, Baghdad in Iraq, Isfahan in Iran, and Samarkand in Uzbekistan. While Europe languished in the Dark Ages (ca. 410–1000 CE), people in the Muslim cities enjoyed fine housing with magnificent carpets, fine furniture, porcelain from China, and the finest foods. The streets were paved; there were running water and sewer systems. Ingenious mills and waterwheels were built along the rivers to provide irrigation. Many advances were made in agriculture, manufacturing, and commerce.

Advances in Education – By the ninth century, paper had been brought in from Asia. Muslim libraries began to fill with manuscripts. Study began in the mosques and eventually moved to colleges called **madrasas**. These Islamic institutions were established a century before the university system was founded in Europe. The Muslims prepared the way by preserving many of the classical disciplines, which became foundational in the European Renaissance education.

The medieval Islamic period produced many excellent scholars. By the ninth century great advances in science, math, medicine, art, music, and education were made. The teaching of the Hadith that "the ink of scholars is more holy than the blood of martyrs" had taken hold, and Islam had developed universities and libraries that were unrivaled. Scholarly contributions came from all parts of the empire: philosophy, mathematics, and astronomy from Greece; the best thinking on technology, agriculture, art, and architecture from Rome, Persia, and North Africa; and the basis for Arabic numerals from India. The technology of the period led Muslims to develop advanced uses of hydropower and wind power.

Unfortunately, much of this learning was lost when the Muslims were driven out of Spain in the thirteenth century during the Crusaders' attempts to capture Cairo, and when the Mongolians sacked and burned Baghdad in 1258.

Scholarship – In the eleventh century an eminent Persian scholar, **Ibn Sina** (*Latin name: Avicenna*) (980–1037), produced works on numerous subjects: medicine, astronomy, chemistry, physics, geology, and mathematics. His work as a physician influenced his later writing on medicine, and his medical texts, *The Canon of Medicine* and *The Book of Healing*, were used widely in the universities of Europe as well as in Asia until the 1650s. Ibn Sina also did extensive interpretations of the works of Aristotle, some of which influenced Thomas Aquinas and other medieval scholars in their application of Aristotle to interpretations of Christian doctrine. His work on the "universal" as a product of the mind and on the function of the intellect influenced many medieval philosophers in Europe.[31] His writings on physics influenced Isaac Newton and Galileo. Several of his encyclopedias were translated into Latin in Venice. In his later years, Ibn Sina turned to the study of literature and language.

Ibn Rushd (*Latin name: Averroes*) (1126–1198) was also learned in a wide variety of disciplines such as music, law, mathematics, geography, physics, and astronomy. His philosophical commentaries attempted to reconcile Aristotelianism with the Islamic faith. When all Muslims, including Ibn Rushd, were driven from Córdoba, Spain, his work found its way into Christian Europe in Hebrew and Latin translations and was most influential on medieval thinkers. His philosophical work, like that of Ibn Sina, provided medieval scholars such as Aquinas access to Aristotle's thinking and provided the philosophical foundation for Catholic scholastic theology until the present day.[32]

Abu Hamed Muhammad ibn Muhammad al-Ghazali (*Latin name: Algazel*) (1058–1111) was one of the most eminent theologians, legalists, and mystics of Sunni Islam. After the completion of his studies at a prestigious madrasa in Gurgan, Persia (now present-day Iran), al-Ghazali became a confidant to the sultan and was appointed to an important professorship in Baghdad. He gained a reputation as the most accomplished and well-established intellectual of his day. In reading Sufi literature, he came to realize that leading an ethical life was not possible while serving the court with all its corruption. Al-Ghazali left his court positions and made a pilgrimage to Jerusalem, where he vowed to avoid serving political authorities. After teaching at a Sufi school, he returned to the prestigious school where he had once studied.

One of the goals of al-Ghazali was to defend Sunni theology against challenges from Shiite Ismaili theology and the Arabic Aristotelian philosophy. His critique of twenty of the Shia positions is considered to be a landmark in the history of philosophy and was extremely influential on later medieval philosophy. Al-Ghazali was able to properly and successfully introduce Aristotelianism into Muslim theology, mainly by showing that there is no contradiction between reason and faith.[33]

Another important contribution of Al-Ghazali was his demonstration that the legalistic shariah could be successfully integrated with the mystical and loving thought of Sufism. This is a milestone in Islamic thought because it demonstrates that one can be a Muslim conforming to the laws of Islam and still be a person devoted to piety and even mysticism. This work is still studied by those who seek direction with their spiritual struggle to be in the world but not of it.

Literature – Muslims take pride in their poetry, which is often recited or sung in public. Here is a sampling:

Abu Nuwas (d. 815), a satirist, pokes fun at his tea-totaling fellow Muslims.

"While the flask goes twinkling round,
Pour me a cup that leaves me drowned with oblivion,
ne'er so nigh,
Let the shrill muezzin cry."[34]

Another poet, Abu al-'Ala' al-Ma'arri (d. 1057) is cynical about religion.
"Now this religion happens to prevail,
Until by that one is overthrown,
Because men dare not live with men alone,
But always with another fairy tale.
A church, a temple, or a Kaaba stone, Koran or Bible or a martyr's bone."[35]

>> **Listen to the reading of verses from The Rubaiyat of Omar Khayyam:**
http://www.youtube.com/watch?v=8eHWU3MI_V4&NR=1

There is a vast body of Arab literature. Two of the best known Arab poets in the West are Omar Khayyam (1048–1131 CE) and Rumi. The *Rubaiyat of Omar Khayyam* was translated into English by Edward Fitzgerald and has become a classic. Rumi (1207–1273 CE), a Sufi poet, is still extremely popular around the world.[36]

>> **Ponder the meaning of the Sufi poem Say I am You**
http://www.youtube.com/watch?v=QqVBGv2hpQ4&feature=related and
along with Rumi Poem, Iranian Music and Divine Dance
http://www.youtube.com/watch?v=hqhNPY882kE
and experience some of Islam's contributions to the arts and literature.

Architecture – Architecture was and is important for Muslims, and though they did not normally allow any images, marvelous designs adorn their edifices. Enormous palace complexes along with magnificent mosques with wonderful domes and towering minarets were built during this Golden Age.

In the seventh century the amazing **Dome of the Rock** was constructed in Jerusalem on the foundation walls of Herod's Temple. Considered by Muslims to be the site of Muhammad's ascension into heaven accompanied by the angel Gabriel, Jews traditionally view it as the place where Abraham prepared to sacrifice his son Isaac. Because of this, it stands today as one of the most controversial buildings for Jews and Muslims.

In the early eighth century, the Umayyad Mosque was built in Damascus, Syria. This mosque is also known as the **Grand Mosque of Damascus**. It is one of the largest and oldest mosques in the world and has both architectural and religious importance. The mosque was built on one of the holiest sites in the old city, the Christian basilica dedicated to John the Baptist. The mosque includes a shrine containing the head of John the Baptist (in Arabic: *Yahya*), a revered prophet for both Christians and Muslims. The tomb of Saladin, the first Sultan of Egypt and Syria and the Muslim leader who led successful campaigns against the Crusaders, can be found along the north wall of the Grand Mosque.

The magnificent **Sultan Ahmed Mosque**, better known as the Blue Mosque in Istanbul, was built in the early seventeenth century and is known for its blue tiles on the interior walls. It was built during the rule of Ahmed I on the site of the palace of the Byzantine emperors, facing the Hagia Sophia, another key mosque in Istanbul. The Blue Mosque is still used for prayer, but it is also a common stop for tourists.

The well-known Taj Mahal is a mausoleum located in Agra, India. It was built in the mid-seventeenth century by Mughal emperor Shah Jahan in memory of his favorite wife, Mumtaz Mahal. This incomparable structure is known for its wonderful symmetry and for its majestic white marble dome.

Amid all this progress there were many conflicts: battles between the Sunni and Shia, civil wars, the rise and fall of Islamic dynasties, and assassinations.

The Christian Crusades began in 1095 CE, and the Crusaders marched through Turkey and took Jerusalem in 1099 CE, slaughtering the Jews and Muslims there. Syria and Palestine were controlled by the Crusaders. Eventually, the Muslims were able to counterattack, and in the twelfth century the Egyptian sultan Saladin began to win back Muslim territory. In 1187 CE, Saladin soundly defeated the Crusaders and reclaimed Jerusalem. In the thirteenth century the Mongols, led by the infamous Genghis Khan, swept down from the north and attacked the Muslim Empire, destroying some of its cities and slaughtering their populations.

MODERN MOVEMENTS IN ISLAM

By the mid-nineteenth century many of the Islamic countries of South Asia were languishing under British colonial control (the East India Company from 1757–1858 and the British Crown from 1858–1947), which dominated their lands, enslaved their people, and strove to Christianize them. Rebellion was in the air. A resulting revolt in India in 1857 was brutally crushed, and many cities were burned to the ground.

At this point, the Indian intellectuals began to develop a compromise—the so-called Modernist position. A wealthy Hindu by the name of Syed Ahmad Khan proposed that Muslims create a new image by uniting Islamic thought with modern European rationalism and science. Modernists proposed that Islamic teachings and morality could evolve, adjust to modern demands, and be separate from government.

This Modernist message was then brought by Jamal ad-Din al-Afghani (d. 1897) to Egypt, a country whose people were completely under the control of colonialists. Al-Afghani sought to make Islamic values more contemporary, rather than simply aping the values of the West. He was joined by Muhammad Abduh (d. 1905), and together they proposed overcoming Western domination through Pan-Islamism, which demanded Muslim solidarity. This movement soon ended due to the national and religious diversity and lack of solidarity among Muslims.

Enter Hasan al-Banna (d. 1949) who came to Cairo in 1923. Here, Britain had developed a version of apartheid in Egypt, with signs on public places that read: "No Arabs." Al-Banna was appalled by the British oppression. He organized the Society of the Muslim Brothers, whose outcry was "Islam is the answer." This movement of Islamization spread to Syria, Jordan, Algeria, Palestine, Sudan, Iran, and Yemen. It confronted Christian proselytizing, Zionism, and the degradation of the Muslim people, as well as the wealthy and luxurious lifestyles of the Arab monarchs. Al-Banna's controversial interpretation of Islam persuaded him that the pursuit of love and justice should be the true goals of his people. His message and powers of organizing were perceived to be dangerous to the ruling powers, leading to al-Banna's assassination in 1949.

Gamal Abdel Nasser led the overthrow of the Egyptian government in 1952. He feigned loyalty to the Muslim Brothers, but it soon became apparent that he had his own dominating agenda in mind. The Brothers were blamed for an assassination attack on Nasser, and their leaders were imprisoned and tortured, and many were executed.

One survivor of this imprisonment was Sayyid Qutb (d. 1966), who was convinced that Muslims must establish their own states where God and the shariah would reign supreme. Qutb believed that Western thought and culture had become toxic for the Muslim world and had to be replaced by a thoroughly Islamic system. Qutb was arrested and then hanged for treason. His fellow Muslim Brothers, however, fled to a country that was booming with oil money and ready to establish a new Islamic Kingdom—Saudi Arabia.

The First Saudi State was established in 1744 in a desert area where a small clan was led by a sheik named Muhammad ibn Saud (d. 1765). Ibn Saud was joined by an itinerant preacher named Muhammad ibn Abd al-Wahhab, a zealot who wanted to purify Islam of medieval interpretations of the Koran, which were considered to be erroneous, as well as superstitions and false devotions to "saints" and to their places of burial. The two united and embarked on an Islamic fundamentalist crusade that came to be known as **Wahhabism**.[37]

Wahhabism was a form of Sunni Islamic fundamentalism that wanted to restore Islam to what was thought to be its purity in Medina under Muhammad. Anything considered to be false piety, foreign influences from other cultures, or false interpretations were to be eliminated. Those who disagreed, such as the Shia and Sufis, were to be destroyed. Saud's son, Muhammad ibn Saud, began a scorched-earth reform against all Muslim innovators. His warriors conquered the Shia city of Karbala and destroyed the markers of graves that were reverenced by the Shiites. In the early nineteenth century, the House of Saud once again destroyed places of pilgrimage that had become devotional places. All books except the Koran were burned, women were required to be veiled and hidden away, and men were required to wear beards. Many Shiites and Sufis were massacred. In 1818 the Egyptian army drove the Wahhabists back to their original territory.

One hundred years later, the British wanted control of the Saudi people and the Persian Gulf. They provided money and arms to the House of Saud to take over the Arabian peninsula. The Saudi army recaptured Mecca and Medina, executed forty thousand men, and imposed Wahhabism on the area—the new Kingdom of Saudi Arabia. Soon after, the Saudis received what they counted as a special blessing from God—oil was discovered. Saudi Arabia became oil rich, and its wealthy sheiks became major players on the world's economic stage. The Muslim Brothers and Wahhabism had found a rich and secure home and now controlled the central pilgrimage place, Mecca, where millions come with their devotion and their money for hajj.

From Saudi Arabia, Islamic fundamentalism spread to Hamas, a Sunni Muslim group in Palestine, and to groups such as the Islamic Jihad. In 1991, during the Persian Gulf War,[38] a new group of Saudi radicals was formed whose name has become infamous in the Western world—**al-Qaeda**. Led by founder Osama Bin Laden, the group adopted Wahhabism and turned on the Saudi royals, whom they considered to be corrupt, immoral Muslims in league with foreign powers. Al-Qaeda vowed to "wipe out" all such leaders and, indeed, any Muslims whose views of the Koran or shariah differed from their views.[39]

Bin Laden, raised as a devout Wahhabi Muslim, was part of a large, wealthy Saudi family. His father had built many buildings as well as the new mosque in Mecca. As a young man, Bin Laden went off to lead a group of Afghans in a fight to overcome the invasion of the Russians who were supporting the Marxist government of Afghanistan. His efforts in this Afghan civil war were supported by money and arms from the United States. The Russians ultimately withdrew from the Afghan conflict.

Bin Laden eventually turned on the United States, in part for their role in the Persian Gulf War. He decided that the Americans had to be driven from Muslim lands. Bin Laden began a reign of terror in the 1990s with attacks that included the bombing of United States embassies in Africa in 1998 and the American naval vessel the *USS Cole* in 2000. On September 11, 2001, he masterminded the attacks on the World Trade Center's twin towers, the Pentagon, and a failed attack on the United States Capitol. After a ten-year search, Bin Laden was discovered and killed in Pakistan in May 2011.

After 9/11, the United States and the world changed dramatically. Security against terrorism became a global concern. These heinous acts of a small, radical fundamentalist group, the strong emotions they evoke, and a lack of understanding of the Islamic culture have left many with false impressions of the foundations of the Muslim faith and its role in today's world.

Modern Movements in Islam – There are a number of currents in the flow of modern Islam. At the extreme, there are radical fundamentalists who insist on strict reform for Muslims, the curtailing of Western influences, and rigid enforcement of shariah law on all aspects of life. This movement is often referred to as Islamism. Some of these fundamentalists revert to terrorism and suicide bombings to make their point and destabilize their enemies. We see these movements in places such as Afghanistan, Iraq, Iran, and Saudi Arabia.

Among many Muslims, especially the young, there is serious unrest with the old ways of fundamentalism and a desire for freedom, democracy, and peaceful normalcy. Demonstrations and rebellions in a number of Arab countries indicate that these movements are growing in places such as Libya, Egypt, Yemen, Syria, and Jordan. Many Muslims realize that Islam once led the way in cultural progress, and they want to once again take their place as leaders. Many resent being identified with terrorism and suicide bombings and firmly believe that these are against the teaching of the Koran. They want to rid themselves of extremists who kill Muslims and non-Muslims in great numbers, oppress other religions, and demand to take control with brutal authority.

Still another trend in Islam is secularism, which looks to adopt modern ways and to limit shariah law from dominating many aspects of life. The classic example is Turkey, which separates state and religion and allows significant cultural freedom for its Muslim citizens. In other Muslim countries, such as Jordan, certain personal freedoms are allowed, but shariah law still controls such areas as marriage, divorce, and inheritance. Growing numbers of Muslims in Europe and the United States have adapted to modern ways of living, yet still maintain their religious traditions.

CONNECTING ISLAM WITH WORLD ISSUES

We have explored the history and teachings of Muhammad, the basic tenets of Islam, its various religious sects and their history and contributions, and the politics of Islam's changing position within the world. The Five Pillars of Faith, rituals that define what it means to be a Muslim, demonstrate how Islam is a religion of action.

Many Muslims are experiencing prejudice and oppression and are therefore caught in a struggle for civil rights and equality in the Western countries where they live. Though citizens of many Western nations, Muslims have little chance to be elected to office and are often deprived of their civil rights. In airports they often find themselves suspect and even shunned. Many are reluctant to wear their traditional dress. In France it is illegal for women to wear the Islamic veil.

In many Muslim countries there is a strong resistance to strict shariah law and a desire for more freedom and democracy. Young people especially want to be part of contemporary culture and free to pursue careers for a fair wage. In the past, the money in many of these countries has gone to the lush lifestyles of the rich dictators, while ordinary people were forced to the economic margins. Access to the Internet and the media has made many modern Muslims often realize that they are living on the margins in antiquated cultures with inadequate opportunities for education and careers. Modern Muslims today often struggle with issues of despotism, inequity, intolerance, and fanaticism.

In the next section we will discuss specific Islamic values that Muslims are connecting to their concerns about the environment, peace, and women's issues. We will also be introduced to Muslim activists who are applying their faith to work in these areas.

>> *Consider the comments of Ramadan Tariq:* Knowledge of Present Realities *at http://www.dailymotion.com/video/x95ztg_ramadan-tariq-knowledge-of-present_webcam* **to hear a modern scholar discuss the role of Islam in the world today.**

Islamic Values and Ecology – The great Sufi poet Rumi seems to have been prophetic of his people's modern care for the earth when he wrote in the thirteenth century:

> *"We began as a mineral. We emerged into plant life and into animal state, and then into human, and always we have forgotten our former states, except in early spring, when we slightly recall being green again."*[40]

Islam has only recently awakened to the role it needs to play in caring for the environment. Fazlun Khalid, an influential Muslim environmentalist, points out that in the mid-1980s, when he first began his work in that field, few Muslims had an interest in the subject.[41]

There are a number of reasons why Muslims have been slow to engage in ecology. First, in the past many Muslim leaders took the position that the environmental crisis was of Western making. Their attitude was summed up as: *They caused it. Let them fix it.* Others firmly believed that from the beginning Islam had all the teachings needed to take care of the earth and that they have always done their part.

In the last few decades there have been significant changes in the Muslim attitudes toward the environment. Many in Islam have come to recognize that they, too, are part of the problem. They acknowledge that many of the oil products that cause damage come from Arab countries. Many Muslim countries now admit that they are part of the industrialization, pollution, consumerism, and waste that contribute to the environmental crisis.

As a result of accepting their environmental responsibilities, many Muslim leaders have stepped out of their "fortress mentality" to become involved in local, national, and worldwide movements for caring for the earth and its resources. In 1983 the International Union for the Conservation of Nature and Natural Resources was founded to set policies for Saudi Arabia and other Muslim countries. Large conferences were held in Iran, Saudi Arabia, and throughout the Middle East. Many grassroots advances have been made in Iran by environmental groups such as the Green Front of Iran, which has full support of the Islamist government. In Turkey, a secular nation where most of the citizens are Muslim, there are more than a dozen influential environmental organizations. In Pakistan, both environmental agencies have combined with the government in protecting coastal ecosystems and water sources and promoting cleaner production standards. In Nigeria, Islamic teachings have been applied to the conservation of water and forests, the preservation of wildlife, and the proper use of resources.

With regard to Islamic teachings, many Muslims have moved beyond the self-assured quoting of "feel good" statements from the Koran and the Hadith to more careful and critical reading of the sacred texts.[42] Islamic teachings are now seen in their context, when the understanding of nature was quite different and there was not an awareness of the dangers of environmental problems. Progressive Islamic scholars are now reconstructing the ancient teachings and searching for new meanings that are applicable to today's ecological issues. Some of their work is reflected in the modern-day Islamic values for ecology of unity, trusteeship, and accountability.

• Unity (Tawheed)

The Koran teaches that the entire universe is from Allah, the One God's creation. Allah's unity is reflected in the oneness of humanity and creation. All creation belongs to Allah. Humans have been created to sustain the unity of creation, to keep balance and harmony in the world, and to avoid any discord or abuse that would disturb the unity of creation.

Indeed, Muhammad's mission was to reveal the need for unity of all things with the One God. All traditional Muslims seek this through prayer and obedience to the law, while the Sufis seek this unity through mystical experience. Others stress that this unity can best be achieved through consensus and collaboration. Islamic environmentalists today often call upon this sacred unity, and they ask Muslims around the world to be concerned with the serious problems of pollution, degradation, and the waste of resources.

>> See Islamic Foundation for Ecology and Environmental Sciences and its statement to the United Nations at the 2009 Climate Change Summit at http://www.youtube.com/watch?v=wwl0pk_f8S0&feature=related for more on how Muslims apply their faith to their environment.

• Trusteeship (Khalifa)

Muslims believe that all creation belongs to Allah and that Allah created humans to act as trusted protectors or guardians of creation. The Koran says, "And We have not sent you but as a Mercy to the worlds."[43]

Humans are not to be masters of the earth for it does not belong to them. They must show the same benevolence and mercy toward creation as does the Creator. Islam teaches that Allah has provided his followers with guidance as to what is harmful and forbidden, what is helpful and approved. People must be guided by these principles in every aspect of private and communal life. These principles command that people make proper use of all the resources given by God, even if this requires sacrifice.

•*Accountability*

Muslims believe that all humans will one day be accountable for their actions. Good actions will be rewarded, and bad actions punished. This applies to how humans carry out their responsibilities toward creation. *Islam* means submission to the responsibilities given by Allah. All humans are called to submit to the responsibilities given them toward the earth and to deal with its resources in a peaceful and benevolent manner, after the example of Allah.

The Koran speaks of civilizations that have been punished for their evil actions:

> *"Do they not learn a lesson from the chronicles of history, and don't they see how many a nation we have made extinct in the past? We had showered on them both celestial bliss and earthly affluence. Yet when they became evil, We toppled them down and raised new civilizations in their wake."*[44]

Muslim ecologists today interpret such sobering passages in terms of the punishments people are bringing upon themselves for their irresponsible degradation of creation.[45]

Muslims in Action

Yuyun Ismawati – Yuyun Ismawati works as an environmental engineer in Indonesia. She has designed water and sewage systems in both Indonesian cities and rural areas. Her programs offer opportunities for employment as well as environmental education for low-income people. She also helps poor communities with waste disposal, providing safety for the children whose health is endangered by fiery garbage dumps. Her programs help empower the poor to improve their environment.

>> *Watch Yuyun Ismawati's video 2009 Goldman Prize for* Islands & Island Nations *where she describes her waste management work in Bali:* *http://www.youtube.com/watch?v=NpNLPFv-QY4.*

Fazlun Khalid – Fazlun Khalid is the founder and director of the **Islamic Foundation for Ecology and Environmental Sciences**. Khalid is one of the most outstanding examples of a Muslim dedicated to the future of the earth. He travels the world working to reeducate Muslims in Islamic environmental values. He says:

"It hasn't quite entered the human consciousness, that if planet earth suffers, we suffer and that we have nowhere else to go. We are part of an integrated earth, and when we reduce the natural world to an exploitable resource this turns inwards on us. How else does one explain the consequences of climate change?"[46]

>> *Listen to Dr. Fazlun Khalid's talk,* Muslims and the Environment Crisis 1-3 *http://www.youtube.com/watch?v=dCKUoXmOQp4 and identify connections between his mission and the Islamic faith.*

Masoumeh Ebtekar – Masoumeh Ebtekar is an Iranian woman who has been vice president of Iran and head of the Environment Protection Organization of Iran. A university professor, she holds a Ph.D. in science. She has worked to show how wars are responsible for environmental devastation. It is her position that war against people is also a war against nature. Ebtekar was given the Champion of the Earth Award by the United Nations in 2006. In 2008 a survey in the British newspaper *Guardian* asked who around the world can bring about change and radical solutions to environmental problems, help people modify their lifestyles, and influence politicians and businesses to be more aware of dangers to the earth and its resources. Dr. Ebtekar placed among the top fifty people. Despite the lack of governmental support, Dr. Ebtek-ar has helped establish thousands of environmental groups led by women seeking change. She says: "We need to put spiritual and ethical values into the political arena...You don't see the power of love, you don't see the power of the spirit, and as long as that goes on, the environment is going to be degraded, and women are going to be in very difficult circumstances."[47]

Islamic Values and Peace – As mentioned earlier, the Western world changed dramatically after September 11, 2001. The United States de-clared war on terrorists, and the world became extremely security con-scious. Airports were transformed and security screenings were inten-sified. More importantly for Muslims, they were often profiled as Arab terrorists. The media emphasized images of bearded men in black, their faces covered, training in the deserts of South Asia, presumably for an-other attack. Arab men (and sometimes women) were shown being tor-tured in prison, or on trial, or blowing themselves up as suicide bombers.

There are 1.3 billion Muslims in the world—comparatively speaking, almost none are terrorists. Nonetheless, Muslims have had to suffer seri-ous discrimination. Their vehement protests that such violence is incom-patible with Muslim teaching have often gone unnoticed. Many scholars, both Muslim and non-Muslim, have clearly demonstrated that terrorism is not justified by Islam, or any religion for that matter. They point out that the word Islam is in fact derived from the word *salaam* meaning peace, and that most Muslims are deeply committed to world peace. At the same time, the majority of Muslims have been struggling against violence and terrorism. In this section we discuss some of the main Islamic values that reject terrorism and cite a number of significant Islamic peacemakers.

• The sacredness of human life
Both the Koran and Hadith affirm the common origin of hu-manity, the equality, and solidarity of all people. A central notion

in Islam is that humans are the most dignified of all creatures. They come from God and return to God. God's message through the Prophet is to show them how to recognize and develop their connection with God and to teach them that creation is in their trust. In the Koran, God says: "Surely the earth belongs to God and He bequeaths it to such of His servants as He pleases."[48] He asked the angels to bow to Adam and gave him and his spouse a garden where they could eat, with the exception of one tree. Satan tempted them; they transgressed and were banished. But God came to them, was compassionate and kind, and promised them guidance. God also says, "We created man of finest possibilities." The Koran also notes that God created diversity: man and woman, different tribes and nations, "that they may recognize one another." They have been made to be brothers and sisters—friends.[49]

For Islam, life is sacred and must be protected and preserved from violence. In the Koran, God says:

"That is why we decreed for the children of Israel that whosoever kills a human being, except for as punishment for murder or for spreading corruption in the land, shall be like killing all humanity; and whosoever saves a life, saves the entire human race."[50]

In battle, Muhammad forbids the killing of women, children, or the aged.

>> **Watch a moving video and listen to the song Try Not to Cry,**
a commentary on the impact of conflict in Palestine on the children.
Free Palestine - *Sami Yusuf*:
http://www.youtube.com/watch?v=QpUC45vugdY

• *Justice*

Muhammad called his people to establish a just social society, free from the old blood feuds, tribal revenge-taking, and repression by dictatorial leaders. The Koran says: "Verily God has enjoined justice, the doing of good, and the giving of gifts to your relatives; and forbidden indecency, impropriety and oppression."[51]

Allah commands his people to be "custodians of justice and witnesses for God, even against yourselves or your parents or your relatives. Whether a man be rich or poor...."[52] The Koran teaches that justice is next to piety. Over the ages, Muslims have relied on the Islamic scholars to determine what revelation and reason teach about specific measures of justice. Today many Muslims around the world engage in the struggle for justice.[53]

• *Jihad*

Jihad is often a frightening word for Westerners; it denotes danger, violence, suicide bombers, and fear. As a matter of fact, the word jihad has many connotations other than war. Jihad appears thirty-five times in the Koran, and only four of those verses are clearly warlike in intention. There are more than seventy places in the Koran where war is prohibited. At the same time, there are a number of verses that can be interpreted either as peaceful or warlike, and these passages have been interpreted by terrorists to justify violence.[54]

The word jihad literally means struggle. The Greater Struggle for Muslims is against the sinful and dark side of human nature. That includes the struggle against greed, violence, and hatred. Jihad promotes the struggle for a more peaceful and just society and the betterment of the human condition. The "Lesser Struggle" involves the use of arms.

Like Buddhism, modern Islam calls for a "middle way" for renewal, one that includes openness, respect, and oneness with all peoples.[55] For many Muslims today this struggle is against terrorism of all kinds, including Muslim against Muslim. It is a struggle to get beyond debates about rituals, length of beards, and head coverings and concentrate on the real-world issues of illiteracy, refugees, poverty, hunger, war, and unemployment. For most Muslims, the key jihad is the inner jihad to strive to lead good, loving, and responsible lives.

Many Muslims want to shed the stereotype of being "people of the sword," who want to spread Islam with military might and terror. The modern Muslim is typically concerned about making ends meet, getting an education, holding down a job, and, if married, raising a family. One out of five people in the world is Muslim, and the vast majority are people who struggle to be good parents, diligent workers, and friendly neighbors.

There is admittedly a warlike jihad in the Koran, and it is similar to the "just war theory" proposed by the Christian theologian Augustine of Hippo (345–430) and used by many Western nations still today. Here, war is to be of a last resort, strictly a defensive effort, and one that protects noncombatants. In no way does this notion of jihad justify wanton acts of terrorism that kill and maim innocent civilians. The vast majority of Muslims are appalled by the views of extremists like Osama Bin Laden and consider their views on jihad to be completely erroneous and dangerous.[56]

>> **Watch CNN Being Muslim in America**
http://www.youtube.com/watch?v=7JPDLkzwN3A
for a discussion of terrorism and how the term is applied in the United States.

Muslims in Action

Anwar al-Bunni – Anwar al-Bunni is a human rights lawyer in Syria who recently completed a five-year prison term for his conflicts with the government. Earlier he had denounced torture and mistreatment in the Syrian prisons, especially in the case of a young man who died in prison and whose body showed signs of torture. He has challenged the government on other occasions for abuse of legal rights and for using torture in its prisons. He has served prison terms for his confrontations with the government, as have four of his siblings. While al-Bunni was in law school, three of his brothers and one sister were imprisoned for political reasons. Through the media, he has kept the world informed of human rights abuse in Syria. As a result, he has been harassed, threatened, kept under surveillance, and jailed. Germany has awarded him its prestigious human rights prize from the German Association of Judges. In 2008 he was awarded the Front Line Award for Human Rights Defenders at Risk, given in Ireland. He was released from prison in 2011.

>> *Learn why the 2008 Front Line Award for Human Rights Defenders at Risk was bestowed upon Anwar al-Bunni of Syria:*
http://www.youtube.com/watch?v=q484_7IPA2A

Shirin Ebadi – Shirin Ebadi has a long history in the struggle for peace and justice in her homeland of Iran. Born in 1947, she grew up in Tehran in a highly educated family. After receiving a law degree from Tehran University, she began work as a judge. After the Iranian revolution of 1979, the Ayatollah Khomeini turned Iran into an Islamic state and took on autocratic powers. Ebadi and the other women judges were dismissed and given clerical jobs. She decided to retire, and being housebound, she opened a private practice and began to write books on social justice. In her law practice she took on extremely controversial cases: one involving some students killed in an attack on university dorms and another of the murder of a photojournalist. She cofounded the **Centre for the Defenders of Human Rights**. Ebadi lectures at the University of Tehran and gives human rights seminars worldwide. She has been a strong advocate for the civil rights of women and children.

In December 2009, security forces raided and shut down the Centre for the Defenders of Human Rights and confiscated documents about her clients, who include some of Iran's most important political figures of the last thirty years. Her alleged declaration supporting the Palestinians in Gaza was the reason given for the closure. Since then, her former secretary was arrested. Right-wing crowds have gathered outside her home, where they accused her of supporting the United States and Israel and vandalized her house.

When she called the police they came and stood by, watching the vandalism and doing nothing to stop it. Ebadi is a strong supporter of two separate states for Israel and Palestine. She vehemently opposes the possession of any nuclear weapons in the Middle East, including India, Pakistan, and Israel. She has vigorously defended those in jail for protesting the corruption in the 2010 Iranian presidential election.

>> *See* **Shirin Ebadi: Iran Awakening**
//www.youtube.com/watch?v=AmW_IyDvalE

and **Conversations with History: Shirin Ebadi**
http://www.youtube.com/watch?v=xDBa44vl-vU to learn more about her work for peace.

Tawakkul Karman – Tawakkul Karman won the Nobel Peace Prize in 2011. She is a mother of three, who was born in a rural area of Yemen in 1979. She is a human rights activist and politician in Yemen, a country recently experiencing a large-scale popular movement against the government. Karman organized student rallies in the capital of Sana'a, after which she was arrested. When she was released, she immediately went back to protesting.

"I am very, very happy about this prize," Karman told the Associated Press. "I give the prize to the youth of revolution in Yemen and the Yemeni people."

Karman holds a B.S. in commerce and an M.A. in political science. She is also a member of the Yemen Parliament and a prominent advocate for free press and human rights. She has been organizing illegal but nonviolent demonstrations against the regime since 2001. She has made documentaries on human rights abuses in Yemen and has spoken out strongly against tribal leaders who steal lands from villagers.

Arsalan Iftikhar – Arsalan Iftikhar is a young human rights lawyer, who refers to himself as "that Muslim guy," a nickname jokingly derived from the inability of CNN correspondents and other journalists to correctly pronounce his name. He received his law degree from Washington University School of Law in 2003. His specialty is human rights law. He is a devout Muslim and a proponent of what he calls "Islamic Pacifism." Arsalan strongly opposes the warmongering of today. He is a follower of Gandhi and post-hajj Malcolm X. He believes that the peace movement should include all races and religions.

Arsalan has a special appeal to youth. He wants Islam to be "cool" and is concerned about bringing the Koran up-to-date with contemporary life. He writes that the Muslim terrorists "have lost their bloody mind and are simply committing irreligious acts of mass murder that have nothing to do with the true faith of Islam." He wants to be part of a billion Muslim pacifist brothers and sisters who challenge the terrorists who attempt to hijack Islam. Arsalan is passionate about taking Islam back from the terrorists, but always through peaceful means. Arsalan is an international human rights lawyer, who is dedicated to representing human rights cases for many people of all nationalities and religions. This striking, young Muslim leader certainly offers a new and refreshing image of Islam for Americans and the world. Beneath his humor and suavity, there seems to be a deep commitment to nonviolence and peace.

>> See *Young Muslims for Peace* **Foster Peace not War** *and their letter to President Obama:* http://payvand.com/news/09/feb/1149.html

Islamic Values and Women's Issues −

>> *View the video* Federation of Muslim Women
http://www.youtube.com/watch?v=xZyAhL8n7yQ
for news from the Federations in various countries.

The majority of Muslim women in the world are poor and illiterate. They are often refugees fleeing from violence and wars, as in Darfur, Palestine, Lebanon, Pakistan, Afghanistan, and Ethiopia. Many are migrant workers in Europe, or victims of drought or AIDS in Africa. Their issues center around survival for themselves and their children: they seek shelter, food, water, health care, and basic education. Muslim women in other parts of the world, especially the Arab world, are also concerned with the issues of equality, civil rights, marriage and divorce, access to higher education, dress, sequestering in the home, and violence against women, including honor killings. In Islamic countries, where the shariah law prevails or at least has much influence, they often find that their rights are severely circumscribed by conservative clerics. This is true in countries such as Afghanistan under the Taliban, Saudi Arabia, Iraq, Iran, Sudan, and Somalia. Muslim women in Asia and Africa often confront patriarchal cultures that see women as weak, inferior, and seductive.

>> *Read* **Nadia, Captive of Hope** *by Fay Afaf Kanafani,*
a classic memoir of a Muslim woman's experience in the Middle East.

In the United States, Muslim women are at times harassed because of their dress, are often stereotyped as being dominated by males, and are commonly associated with Arab terrorism. Such stereotyping ignores the fact that many American Muslim women function as doctors, lawyers, executives, professors, social workers, U.S. combat troops, and even fighter pilots.

In short, Muslim women around the world are engaged in a struggle (gender jihad) to be treated equally and justly within their own religion and in the cultures where they live.

It is ironic that modern Muslim feminism began with male Egyptian judge Qasim Amin. In 1899 Amin wrote *The Liberation of Women*, a book that ignited a firestorm of controversy. Amin argued that if women were emancipated in Egypt the way there were in the Western world, then Egypt would be strengthened and British colonial rule could be thrown off.

Modern Muslim feminists have moved beyond Amin's Western model of liberation and focus on Muslim women's own unique reflections and experiences. Their champion is an Egyptian woman, Huda Shaarawi (1879–1947), a charismatic leader who championed the liberation of Pan-Arab women from the discrimination they experience in their individual countries.

Today there is a strong Muslim feminist movement that endeavors to provide education in the Islamic tradition. Muslim women are well aware that the transmission and interpretation of the tradition has, for the most part, been a male endeavor. They know that the time has come for feminine perspectives and interpretations that arise out of the contemporary experience of women. Many women scholars and activists want to rediscover the radically liberating revelations of Muhammad, sort out those notions that still reflect the cultural limitations of his time, and set aside the many misogynistic interpretations of the Islamic traditions that have accumulated over time. Knowledge is power, and many Muslim women want the power to liberate themselves and live as equals in just societies.

>> *Watch the video* **Tariq Ramadan on Islamic Feminism**
http://www.youtube.com/watch?v=Do--YdH-888
and reflect on what it means to be a Muslim woman today.

In the following section, we will discuss **key Muslim values** connected with women's freedom and introduce outstanding individuals who are active in promoting these values.

• Equality

The equality of women and men seems to be at the core of the Koran. In their creation as well as in the final reward they receive, there is no distinction or hierarchy between women and men in the Koran. Women and men come from the same source and are equally watched over by the same benevolent and ever-merciful Allah (who is not described as male). The Koran says:

> "Fear your Lord who created you from a single cell, and from it created its mate, and from the two of them dispersed men and women in multitudes. So fear God in whose name we ask of one another relationships. God surely keeps watch over you."[57]

Many passages in the Koran speak of "humankind," with no distinction between male and female. For many women Muslims, such revelation of equality stands as the bedrock of Islamic revelation. If such equality and divine protection is revealed in Islam, many women point out today, how can there be any toleration for abuse or injustice toward women?

The Koran also teaches that both women and men receive their just rewards or punishments for the lives they lead. Both are enjoined to know that "God is aware of all you do. God has made a promise of forgiveness and the highest reward to those who believe and do good deeds. But those who disbelieve and deny Our revelations are the people of Hell."[58]

As for the traditional "rib" myth, which in Islam, Judaism, and Christianity was used to give women a derivative and secondary position, the "rib" story is not found in the Koran, although it does appear in later Islamic tradition. Nor does the Koran speak of the Fall in the garden, with Eve as the temptress. In addition, the problematic customs of veiling, secluding women, requiring women to serve men at home, separating them at prayers, and stoning women for adultery are not in the Koran. These teachings have their source in leaders such as Umar, a Companion of the Prophet, and not in the Prophet himself.[59]

Muhammad treated women with great respect, consulted them, and appointed them as spiritual guides. His biographers record that Muhammad helped with domestic chores, prayed side by side with women, and gave them a dignity they had never known in pre-Islamic times.[60] Muslim scholars point out that the presence of many women and slaves in his harem has to be attributed to the culture of the time and not to the true revelation of Allah.

• *Justice*

Muhammad showed great respect for women and outlawed the killing of infant females. The Koran says: "He creates whatsoever He wills, bestows daughters to whosoever He will, and gives sons to whom He choose."[61]

Islamic feminists are critical of many Muslim laws, which they claim have no connection with the Koran. Feminist scholars declare many practices to be non-Islamic: the stoning of a woman for adultery, which is still practiced in some countries; requiring a woman to be sequestered in the home; forcing women to cover their entire body when in public; not allowing women to drive cars, still practiced in Saudi Arabia; or forbidding women from marrying non-Muslims. Feminists protest many other Muslim practices for not being in the inspired tradition. In some countries four men must testify in court as witnesses before a woman can prove that she was raped. Forced marriages and "honor killing" of women who resist these marriages or are dishonored in some other way are still common in Muslim communities, even in England.

There are many other issues to which Muslim women object that are actually based on the Koran: polygamy, which allows a man to have four wives; the lesser weight given a woman's testimony in court; a woman receiving less of the share of an inheritance than a man; and unjust divorce laws favoring the husband. Women scholars attribute these laws to the context of culture in Muhammad's time and hold that they have no relevance in today's world. Fundamentalists and many conservative imams, on the other hand, hold that these practices are based on God's own words and cannot be changed.

• *Gender Jihad*

As mentioned earlier, jihad primarily means to struggle against our evil drives and to work for equality and justice. Many women in Islam today find themselves in extremely difficult and sometimes dangerous situations when they try to liberate women from violence, abuse, and helplessness in the face of forced marriages and injustices in divorce. Many Muslim women want to be free to pray and even lead prayer in the mosques, be leaders in their families, dress modestly, but as they please, and have their day in court as equals. They want to be able to single out terrorists, suicide bombers, and extremists for what they are—aberrations from the Islamic tradition rather than faithful followers. They look to some of the noble Muslim women of the past: Khadijah, Muhammad's first wife, who was an independent business woman of means, and who had a profound influence on the Prophet; his later and favorite wife, Aisha, who counseled the Prophet and worked for his mission long after his death as an expert in the Law. They reject patriarchy and male-dominated societies where women are deprived of leadership in the Muslim community and society. They struggle in the name of Allah for the dignity and respect given them in the revelations of the Prophet.

>> **See Muslim Women's League**
http://www.mwlusa.org
for discussions on many Muslim women's issues.

Muslims in Action

Nawal El Saadawi – Nawal El Saadawi is a distinguished and much-awarded Egyptian psychiatrist and author. Her novels and books on women's issues have deeply affected the women's movement in Islam. Her efforts to liberate women have caused her to be fired, banned, imprisoned, and put on death lists. She has been labeled an apostate and was nearly forced by the courts to divorce her own husband as a punishment for her writings and efforts to expose the abuse of Muslim women.

>> *See* **Entrevista a Nawal al-Sadaawi**
> *http://www.youtube.com/watch?v=bmttyRj3NVE&feature=fvw*
> **for insights into this unique personality.**

Amina Wadud, Ph.D. – Amina Wadud was born in 1952 in Maryland. She is a feminist and Koran scholar. She taught for many years at the Virginia Commonwealth University and now teaches at a university in Indonesia. She has led mixed Friday services, as well as highly controversial all-women services. Wadud has received death threats and has been disowned by many conservative imams. Her books *Qu'ran and Woman* (1999) and *Inside the Gender Jihad* (2006) have been controversial, but also extremely influential worldwide among Muslim women. The documentary *The Noble Struggle of Amina Wadud* chronicles her life story.

>> *Watch the video:* **Amina Wadud: Bringing Women's Voices to Sacred Text** *http://www.youtube.com/watch?v=luyjABV1nYw*

Lily Munir – Lily Munir was born in Indonesia and directs the Center for Pesantren and Democracy Studies in Jakarta, Indonesia. Under her leadership, there are nearly twenty thousand schools training the 3 million youth of Indonesia. In her curriculum work, Munir challenges traditional gender norms and teaches that the Koran advocates the empowerment of women, not their domination. She runs seminars on women's health, gender equality, and sexual abuse. Lily Munir is also vigilant that her schools don't become incubators for suicide bombers. She is a strong human rights activist in Indonesia, which has the largest Muslim population of any country in the world.

>> **See Lily Munir:** http://www.youtube.com/watch?v=LXFoJfHPLso

Amel Grami – Amel Grami is a professor in Tunisia and a leading figure in gender jihad. She is an advocate of women's rights to lead prayer and to receive equitable inheritances. It is her position that all the crises in the world and all the pressures put on Muslims gave the sheiks and imams an opportunity to offer young people a secure place to hide in traditional Islam. She sees many of the young people in Tunisia as returning to the past because it is simple and safe. With the banning of liberal feminist books and pressure from the Arab media to be conservative, the young are drawn to old rituals and superstitions. Many young women feel that feminism has actually caused them to lose rights. They want to return to the kitchen and the scarf in order to feel safe. Grami thinks that such narrowness prevents young people from being comfortable with diversity and often moves them toward intolerance or even hatred of any tradition other than their own. She opposes many of the Hadith laws that are being imposed on women as being contradictory to the Koran.

SUMMARY

Islam began in the seventh century CE with the extraordinary revelations of a young organizer of caravans, Muhammad. He became the great Prophet of Islam and used his revelations to revolutionize his area of the Arabian desert. Muhammad's teaching was not acceptable to most in his city of Mecca, so he was forced to withdraw to Medina. There he built a small mosque and led a group of disciples in his new religious reform. Eventually Muhammad was able to conquer Mecca, destroy the idols in the Kaaba, and see to it that only the one God, Allah, prevailed.

Muhammad died in Medina soon after his conquest of Mecca. His revelations were soon gathered by his disciples into the inspired Koran. His values, practices, and laws were collected into the Hadith. The laws, or shariah, became the basis for Muslim life. The Five Pillars of Faith were set upon the foundation of Islam. They are concerned with the profession of faith, prayer, alms, fasting, and pilgrimage to Mecca.

Soon after Muhammad's death, the Muslims conquered most of the Middle East, the Byzantine and Persian empires, and large areas of Europe. From the outset, Islam was divided over the succession of leadership. The Sunnis held that this was to be by appointment, as was the case with the first caliph, Abu Bakr, while the Shiites maintained that this should be by family bonds, as in the case of Ali, Muhammad's cousin and son-in-law. The Kharijites, another radical group, demanded great purity of their leaders. The Sufis represented a sect devoted to mysticism.

During the Dark Ages in Europe, the Islamic Empire flourished into a Golden Age and made great advances in education, science, medicine, mathematics, literature, and architecture. This period went into decline under the pressures of the Crusades, Mongolian invasions, and colonialism.

In the modern era, much of Islam languished under colonial domination. Some Muslim leaders proposed accommodations with Western culture, but more radical groups wanted rebellion. Groups such as the Muslim Brotherhood were founded to purify Islam. Wahhabism was a strong fundamentalist group that settled in Saudi Arabia, where a kingdom was formed that controlled both Mecca and Medina. Out of this movement, another group of Saudi radicals called al-Qaeda was founded in the early 1990s, led by Osama Bin Laden. His group turned on the Saudi royals in order to reform them. He then turned to Afghanistan and, with aid from the United States, fought against the Russian occupiers. Bin Laden then turned on the United States and began to orchestrate terrorist acts, which culminated in the horrendous attacks on September 11, 2001.

Since 9/11, the Muslim world has been struggling with assuring the world that terrorism and suicide bombing are against the Islamic tradition. The killing of Bin Laden in 2011 has given many Muslims hope that they can, in time, put the terrorist image behind them.

In November 2004, King Abdullah II of Jordan and senior Islamic scholars issued a statement that seems to capture the aspirations of many Muslims today. Known as the Amman Message, it describes what Islam is and what it is not, as well as what actions do or do not represent it. The Jordanian king and Islamic scholars point out that the true message of Islam is one of tolerance, brotherhood, and sisterhood—one that embraces all human life and upholds goodness, acceptance, and the honor of all human persons. Islam stands for unity, equality, peace, security, honoring pledges, respect of others and their belongings and property.

In a striking statement, the Amman Message notes that the origin of divine religions is one, and that Muslims believe in all messengers of God and accept their messages. Tolerance and forgiveness are urged, and there is a call to stop fighting non-combatants and to protect innocent civilians. Extremism and the contemporary concept of terrorism are strongly denounced; war is supported only by necessity. The Prophet's teaching is proclaimed: "Take not life which God made sacred." It is the King's hope that this crucial truth about Islam be broadcast to the world.

>> *Contact Muslim youth worldwide at* http://mideastyouth.com *and share thoughts, view videos and podcasts, read articles, and participate in projects.*

CHAPTER 6 VOCABULARY

ayatollah - *Shia religious leader*

fatwa - *Formal legal decision*

hadith - *Collection of sayings by Muhammad*

hajj - *Pilgrimage to Mecca*

hijab - *Head covering for Muslim women*

imam - *Prayer leader*

Islam - *Surrender to God's will*

jihad - *A struggle against evil*

muezzin - *One who calls the Muslims to prayer*

mullah - *One who takes care of the mosque*

Ramadan - *The month of fasting*

salaam - *Peace greeting*

salat - *Prayer*

shahada - *Declaration of faith in the one God, Allah, and his prophet Muhammad*

shariah - *Law based on the Koran*

ummah - *The Muslim community*

zakat - *The alms to be given to the poor*

TEST YOUR LEARNING

1.What justification does Islam have for saying that the Koran is the speech of God?

2. Compare and contrast Muhammad with other founders of religions.

3. Why do you think the Five Pillars of Faith are so important for Muslims?

4. From what you have learned do you think terrorists are justified in the use of jihad to kill others?

5. Which Islamic values might help followers be interested in ecology, in peacemaking, and in women's equality?

APPLYING ISLAM TO WORLD ISSUES

1. Write a brief essay on the positive contributions Islam makes to religious understanding.

2. Cite and elaborate on parallels between the Five Pillars of Faith and Christian values.

SUGGESTED READINGS

Ahmed, Akbar. Islam Today. London: Tauris, 2002.

Esposito, John L. What Everyone Needs to Know About Islam. New York: Oxford University Press, 2002.

Foltz, Richard C. and others, eds. Islam and Ecology. Cambridge, MA: Harvard University Press, 2003.

Hotaling, Ed. Islam Without Illusions. Syracuse: Syracuse University Press, 2003.

Kaltner, John. Islam. Minneapolis: Fortress Press, 2003.

Milton-Edwards, Beverley, Islam and Violence in the Modern Era. New York: Palgrave Macmillan, 2006.

Nasr, Seyyed Hossei. The Heart of Islam. San Francisco: Harper San Francisco, 2002.

Sfeir, Antoine, ed. The Columbia World Dictionary of Islamism. New York: Columbia University Press, 2007.

Shepard, William. Introducing Islam. New York: Routledge, 2009.

Van Doorn-Harder, Pieternella. Women Shaping Islam. Chicago: University of Illinois Press, 2006.

Yazbeck Haddad, Yvonne and others, Muslim Women in America. New York: Oxford University Press, 2006.

VIDEOS

Esposito, John L. Great World Religions: Islam. Chantilly, VA: The Teaching Co., 2003.

Harrison, Ted, dir. Essentials of Faith. Part 3 Princeton, NJ: Pilgrim Productions, 2006.

North South Productions for Channel Four Schools Television. Islam: Sacrifice to Allah. South Charleston, WV : Cambridge Educational; Princeton, N.J.: Films for the Humanities [distributor], 1996.

SWR and Deutsche Welle TransTel, producers. Islam. Princeton, NJ : Films for the Humanities & Sciences, 2001.

NOTES

[1] *Farid Esack,* The Qur'an *(Oxford, England: Oneworld Pub., 2005), 36.*

[2] *Reza Aslan,* No god but God *(New York: Random House, 2006), 16-17.*

[3] *Some sources say "Recite."*

[4] *Also translated: "a clot of blood."*

[5] *See Esack, 39-40. See John L. Esposito,* Islam *(New York: Oxford University Press, 1998), Ch. 1.*

[6] *Karen Armstrong,* Muhammad *(New York: Harper, 2007), 35.*

[7] *Koran, 68:1-70.*

[8] *Koran, 18:110.*

[9] *William Shepard,* Introducing Islam *(New York: Routledge, 2009), 56.*

[10] *Koran, 17:82.*

[11] *Jonathan P. Berkey,* Formation of Islam *(New York: Cambridge University Press, 2003), 65.*

[12] *See Koran, 7:157.*

[13] *Koran, 4:171.*

[14] *Shepard, 73.*

[15] *See Jonathan Berkey,* The Formation of Islam *(New York: Cambridge University Press, 2003), 146-51.*

[16] *Shepard, 59ff.*

[17] *Malise Ruthven,* Islam in the World *(New York: Oxford University Press, 2006), 61.*

[18] *John Renard,* 101 Questions and Answers on Islam *(New York: Gramercy Books, 1998), 73.*

[19] *See Reem Al Faisal,* Hajj *(Reading, UK: Garnet, 2009).*

[20] *Bernard Lewis,* The Arabs in History *(New York: Oxford University Press, 2002), 89ff.*

[21] *Berkey, 84.*

[22] *Jonathan Bloom and Sheila Blair,* Islam *(New Haven, CT: Yale University Press, 2002), 67.*

[23] *Shepard, 46.*

[24] *Bloom, 53.*

[25] *See Heinz Halm, trans. Allison Brown,* The Shi'ites: A Short History *(Princeton: Markus Wiener Publishers, 2007).*

[26] *See Deborah Amos,* Eclipse of the Sunnis: Power, Exile, and Upheaval in the Middle East *(New York: PublicAffairs, 2010).*

[27] *Aslan, 33.*

[28] *Koran, 50:16.*

[29] *Ahmet T. Karamustafa,* Sufism *(Los Angeles: University of California Press, 2007), 19ff.*

[30] *See Martin Lings,* What is Sufism? *(Cambridge, UK: Islamic Text Society, 1999).*

[31] *See L.E. Goodman,* Avicenna *(New York: Routledge, 1992), 123ff.*

[32] *See Oliver Leaman,* An Introduction to Classical Islamic Philosophy *(New York: Cambridge University Press, 2002), 225-237.*

[33] *Avital Wohlmand, trans. David Burrel,* Al-Ghazali, Averroës and the Interpretation of the Qur'an: Common Sense and Philosophy in Islam *(London: Taylor & Francis Group, 2010), 115ff.*

[34] *Shepard, 187.*

[35] *Ibid.*

[36] *Coleman Barks,* The Illuminated Rumi *(New York: Dell, 1997).*

[37] *See Mohammed Ayoob and Hasan Kosebalaban, eds.,* Religion and Politics in Saudi Arabia: Wahhabism and the State *(Boulder, Colorado: Lynne Rienner Publishers, 2009), 57ff.*

[38] *At this time (1990–1991) the international community attempted to take back oil-rich Kuwait from Saddam Hussein of Iraq who had invaded and annexed it in 1990.*

[39] *Simon Reeve,* The New Jackals: Ramzi Yousef, Osama Bin Laden and the Future of Terrorism *(Boston: Northeastern University Press, 2002), 178ff.*

[40] See http://peacefulrivers.homestead.com/Rumipoetry2.html.

[41] Richard C. Foltz and others, eds., Islam and Ecology (Cambridge, MA: Harvard University Press, 2003), 299ff.

[42] Ibid., 100.

[43] Koran, 21:107.

[44] Koran, 6:6.

[45] Fazlun Khalid and Joanne O'Brien, eds., Islam and Ecology (New York: Cassell Pub., 1992), 8.

[46] See UN Climate Change Summit 22 September 2009 Statement at http://www.un.org/sg/statements/?nid=4083.

[47] See http://www.guardian.co.uk/environment/2008/jan/05/activists.ethicalliving.

[48] Koran, 30–37.

[49] Koran, 7:129; 95:4; 49:13.

[50] Koran, 5:32

[51] Koran, 16:90.

[52] Koran, 4:134.

[53] Mohammed Abu-Nimer, Nonviolence and Peace Building in Islam (Miami: University Press of Florida, 2003), 49ff.

[54] Richard Bonney, Jihad: From Qur'an to Bin Laden (New York: Palgrave Macmillan, 2004), 8ff.

[55] Ibrahim M. Abu-Rabi, Intellectual Origins of Islamic Resurgence in the Modern Arab World (Albany, NY: State University of NY Press, 1996), 263.

[56] Bonney, 362.

[57] Koran, 4:10.

[58] Koran, 5:8–10.

[59] Aslan, 71.

[60] Ibid.,150

[61] Koran, 42:49.

CHAPTER 7

Interfaith Dialogue

Pierre Teilhard de Chardin (d. 1955), a renowned scientist and religious visionary, described the movement of evolution as one of divergence and convergence (spreading out and coming together). One anthropological theory suggests that humankind moved out of Africa and spread out through Europe, the Middle East, Asia, and then through the Americas. As humans moved throughout the world, they developed various religions.

In the so-called Axial period around 500 BCE there was a virtual eruption of religions, philosophies, and great thinkers: Confucius and Lao Tzu in China; Hinduism, Buddhism, and Jainism in India; Zoroastrianism in Persia; Shinto in Japan; the great prophets of Israel; Socrates, Plato, and Aristotle in Greece. Most of these religions and philosophies eventually extended to other areas. Five hundred years later Christianity exploded on the scene and proliferated quickly. Six hundred years later Islam appeared and spread like a wildfire throughout much of the known world.

Along with this expansion and growth there was division, both within the religions and among the religions. Most religions eventually divided into a number of sects and churches, for instance: Vaishnavite and Shaivite Hinduism; Theravada and Mahayana Buddhism; Orthodox, Reform, and Conservative Judaism; Orthodox, Protestant, and Catholic Christianity.

Conflict among religions has been common. The Israelites fought with the tribes and nations surrounding their territory. In early Christianity there was conflict between Jews and Christians, as well as between Christians and Gentiles. Christians opposed Muslims and Jews in the Crusades; Hindus fought Muslims and Sikhs in India.

In modern times, it has been Muslims vs. Christians in Sudan, Christians vs. Muslims in Bosnia, Hindus vs. Buddhists in Sri Lanka, and Jews vs. Muslims in Israel. Since the 9/11 attack on various targets in the United States, the rise of al-Qaeda, and the influx of Muslims in Western Europe, there has been much tension between Muslims and some non-Muslims.

None of this speaks well for religions, which preach love, compassion, and peace. Perhaps that is why many religious leaders urgently advocate a coming together of religions to better understand each other and to devise ways that they can work together for peace and justice.

Ewert Cousins, Ph.D. (d. 2009), a well-known expert in religions, maintained that we are going through a second Axial period, which marks a global consciousness on the part of the entire human community. Cousins pointed out that this consciousness is rooted in the earth, as well as in secular concerns that are political, social, and ecological. He noted that during this period interreligious dialogue is not a luxury, but a necessity and must be grounded in spiritual energy if the earth and humanity are to survive. One scholar puts it bluntly: "Death or Dialogue."[1] Another writes: "The image of the earth moving peacefully through space can be the symbol of interreligious dialogue and our common spiritual journey as we move into the global consciousness of the twenty-first century."[2]

In modern history, there have been significant signs that such dialogue is beginning to take root. As early as the nineteenth century, Christian missionary groups began to have meetings that considered the legitimacy of other religions. During the two World Wars, people from differing religions fought next to each other and learned of each other's faiths.

The Council for a Parliament of the World's Religions first met in 1893 in Chicago, Illinois, where leaders of the world's religions met to discuss their concerns. Since 1993, the Parliament has met every five years in some major city. The last meeting was in 2009 in Melbourne, Australia, where five thousand people from eighty nations gathered for dialogue and prayers. Their agenda included extremely relevant issues including world poverty, global warming, forced migration, and honoring indigenous communities.

In 1948 the World Council of Churches was formed and has since been a force for religious dialogue and unity. Composed of many churches and congregations, it has brought about consensus on many doctrinal and theological issues. The WCC has also brought many Christians together to address political and social justice issues.

The Catholic Second Vatican Council (1962–1965) recognized that truth and salvation exist in all religions and declared that to be saved one must simply live a good life and sincerely search for God. The Council strongly advocated dialogue among churches and religions. After the Council, such exchange dialogue became very important—among experts as well as among the laity in many churches and religions. Pope Paul VI was a leader in interfaith dialogue. In 1964 he established the Secretariat for Non-Christian Religions in Rome to link with other religions. He also attended an historic meeting with Orthodox Patriarch Athenagoras in 1964 at the Jordan River in Palestine. Pope John Paul II (papacy: 1978–2005) was also a strong advocate for religious unity. Among many interfaith events, he preached in a Lutheran church, walked down the aisle of the Canterbury Cathedral as an equal with the Archbishop of Canterbury, prayed at the Jewish Western Wall in Jerusalem, condemned anti-Semitism, and led many interfaith gatherings. Perhaps the most dramatic was his invitation of 120 religious leaders (including an African Voodoo priest) to Assisi in 1986 for joint sharing and prayer. Pope Benedict XVI has also made serious efforts to cordially meet with the leaders of other religions. His joint prayer with Muslim leaders in the Blue Mosque in Turkey was most significant.

Other religious leaders have contributed mightily to religious dialogue and unity. The Hindu spiritual leader Gandhi was a world figure who respected all religions, gathered people of different faiths to live together in his ashrams, and strongly advocated peace between Muslims and Hindus, as well as among all world religions. The Tibetan Buddhist leader, the Dalai Lama, has tirelessly toured the world, meeting religious leaders and advocating peace and religious unity. Jewish leaders such as Rabbi Abraham Heschel (1907–1972), a leading Jewish thinker whose family died at the hands of the Nazis, marched with Martin Luther King Jr. at Selma and engaged in many interfaith dialogues. David Wolpe, the present rabbi of the Sinai Temple in Los Angeles, has led significant, although controversial, dialogues with other religions. The World Jewish Congress has also promoted such dialogues and has met with Pope Benedict. The Buddhists Thich Nhat Hanh and Maha Ghosananda have been major figures in interreligious dialogue. Leaders of the Sikh and of the Baha'i traditions have also been eager to participate in interreligious dialogue and to work for world peace.

WHAT IS INTERFAITH DIALOGUE?

Before we discuss what interfaith dialogue is, we might consider what it is not. First, such dialogue should not be a debate, where two people are trying to convince the other that each is wrong. We should not come to interfaith dialogue with the notion that our religion is the only correct one.

Secondly, interfaith dialogue should not be a missionary activity where we are trying to convert others to our religion. This kind of dialogue belongs in mission work, catechumenate studies, or recruiting sessions, where people come to inquire about joining one faith or another.

Nor should this be an activity where differences are ignored or where there is an effort to join all the religions into one big religion. Such an approach ignores the unique and diverse beliefs of each religion. While parallels and similarities will be discovered, religions have always been different and will always remain so.

Authentic interreligious dialogue is a mutual sharing of beliefs, values, and rituals. It is a sincere sharing of religious journeys and searching, where honest and trusting relationships are established. It is a process where mutual respect, understanding, and enrichment are established. It is a time to share one's faith life and one's religious concerns about world issues; it is an opportunity to explore how religions might join together in actions that will make the world better.

Interreligious dialogue recognizes the differences among religions. Different images or metaphors have been used to accept these differences and yet still be able to relate and dialogue with those who have other beliefs. Some compare this to climbing a mountain by taking differing paths and not being sure what is at the summit. Others compare this to different individuals pointing their fingers to the same moon: perhaps following different rays to the same light source. We might look at religious dialogue as though we are taking pictures of the sun with others at different times of the day. The sun remains the same, but the cameras are capturing it from so many different perspectives. Possibly, we might see ourselves joined with others in carrying parts of the same truth, but in different vessels.[3]

Interreligious dialogue is stepping out of our comfort zone, our "bubble," and trying to understand what another believes. It is never easy!

Interfaith Dialogue: Uncertain Beginnings

I remember one humorous story when I took a group of adult students to the local Hindu temple for a worship service. One woman paused as we went up the steps and said: "I don't think I should be doing this. I was always taught that it was sinful to participate in false religions." I assured her that we were only going into the temple to observe and not necessarily to participate. The woman decided to go in the temple. As the singing and dancing progressed I could see that the woman was being drawn into the ritual. Then I saw her accept a tray with candles and begin gesturing before one of the goddesses! Afterwards, she seemed quite pleased with herself and remarked that the whole experience was very moving for her!

Four Elements Of Dialogue – Paul Knitter, an expert in interreligious dialogue, points out that there are four pivotal elements in such dialogue. First, **the willingness to face and experience the differences**. If we come in thinking that all religions are the same, we probably are in for a shock. There is much difference among the notions of God or gods in the Hindu, Shinto, and Jewish religions—not to mention that Theravada Buddhism does not mention God at all. And what a contrast between the Hindu and Buddhist beliefs in rebirth, the Christian belief in afterlife, and the common Jewish belief that we live on in memory or achievements.

The second element is **putting our trust in the notion that these differences can bring us together rather than divide us**. We can agree to disagree, be enriched by new perspectives about our own religion as well as about the beliefs of others. We can see that all are searching for "ultimate mystery" and that such mystery is in fact beyond human comprehension, and only partially represented in each religion. We can see ourselves engaged in a common search for truth, but knowing that we all "see through a glass, darkly" (1 Corinthians 13:12).

The third element is **witnessing**. This does not mean that we are trying to convert our dialogue partner, but rather we are trying to honestly share our search, our values, our doubts, and religious experiences. We may both find that we are truly inspired by the other's testimony, even changed by it. At the same time, it should be clear that no one needs to reveal anything that feels uncomfortable or too personal. Nor should pressure be put on others to share more than they are willing.

And the final element is **the resolution to carefully listen, to be open to new insights, and even to be willing to change one's religious perspectives**. Often, when we listen to another's view of our religion, we are able to clear up misunderstandings. On the other hand, we might be brought to notice something in our tradition that might truly be wrong and be moved to correct it in our outlook and practice. For instance, Catholics sometimes discover that they can be arrogant and look down on other religions. Dialogue might help such people make an attitude check.[4]

Four Types Of Dialogue – It has been suggested that there are four types of interreligious dialogue: **a dialogue of life**, where we share in a trusting and friendly manner our values, concerns, setbacks, and successes in the context of faith (it might be described as "telling our story"); **a dialogue of action**, where we partner with others in actions to promote peace and justice; **a dialogue of theological exchange,** where we share our interpretation and understanding of religious beliefs; and **a dialogue of religious experience**, where we share religious rituals, prayers, or methods of meditation.[5]

Why Dialogue?

There are a number of reasons for interreligious dialogue. Among the many goals that can be achieved by such engagement are:

Self-exploration

Interreligious dialogue can benefit us by helping us explore our search for beliefs and values. Many college students go on what might be called a "religious leave of absence," in order to assess whether or not they accept the religious tradition of their family. College is often a time to think through one's beliefs and examine the beliefs of others to see where one stands. Interreligious dialogue provides an occasion for expressing what we believe (or don't believe) and at the same time come to understand the beliefs of others. It's an occasion for seeing parallels with other faiths, contradictions, as well as utterly new perspectives—and then letting this enrich our religious perspectives.

An understanding of the faith of others

This type of dialogue is more than just listening to the facts of another religion. It also includes carefully listening to how the other person interprets and relates to his or her religion (or to his or her opposition to religion). It is an opportunity to listen to another's search for purpose in life and ultimate meaning. We can gain understanding of another's deepest questions, empathy for a person's doubts (which might be similar to our own), and a deeper respect for the mystery in religion.

Preparing for a global future

We all know that the world is now smaller because of the ease in travel and communication. Air travel today means that in a matter of hours we can be in an entirely different region with languages, culture, and religions that are unfamiliar to us. In our colleges today we see people from many different cultures and religions coming together to learn. The Internet can instantly put us in touch with diverse people both locally and around the world. Economies and business today are global. This reality puts demands on all of us to learn other languages, be comfortable in other cultures, and be familiar with other religions. Dealing with all of this puts pressure on us to move out of our cultural "bubbles" and become familiar with and respect cultural and religious differences. Thus the need for serious interfaith dialogue in today's world.

Gearing up to make a difference

As we share our ideals and values and become more aware of the needs of others, we can be drawn into action for others. Today we see students going to underdeveloped areas in their own country and abroad to pitch in with a helping hand. Students who share the same faith, as well as students who have differing faiths, are teaming up to build houses in Kentucky, build schools in Africa, work in orphanages in India, serve food in Peru, or work in homeless shelters in Nicaragua. Working together, partying together, and praying together, these armies of young people are becoming a force in the world.

HOW TO DIALOGUE

Here we offer a number of suggestions for productive interreligious dialogue. First, let us consider the personal dispositions that are useful in such dialogue. All those involved need to come with attitudes of equality and respect. A "holier than thou" or "my religion is better than yours" attitude militates against good dialogue.

Experts recommend that we come to interreligious dialogue with an attitude of inclusivity, rather than exclusivity. Neither the exclusive Protestant attitude that one must accept Jesus to be saved, nor the Catholic view that "outside the church there is no salvation or that Protestant churches are "defective," is conducive to productive dialogue.

An acceptance of pluralism is also important in interreligious dialogue, meaning that we accept that there are many authentic religions with their truth claims and the capacity to lead their constituents to ultimacy. This final goal might be the understanding of such ultimate terms as enlightenment, nirvana, eternal life, union with God, or liberation.[6]

Such dialogue should be approached with humility, honesty, sincerity, openness, the ability to "beg to disagree." This is a sharing between equals who trust each other and speak to each other as "I to thou" and not "I to it," to use the words of Martin Buber. All participants are willing to be critical of their tradition and eager to learn about other traditions from those within that tradition.[7]

Most certainly all participants can remain firm in their beliefs, but this does not rule out the willingness to listen to and try to understand and experience another religious tradition.[8] At the same time, there can be an openness to change one's mind about other religions' beliefs, possibly to even accept one or the other of these beliefs.

It is useful to begin with personal stories about one's religious upbringing, influences, and experiences. This helps build relationships of trust and respect, a solid foundation upon which to build a productive dialogue. It is also useful to start by visiting churches, temples, and synagogues to experience how the various religions gather and worship. This way we can perhaps experience the religion rather than merely hearing ideas.

At the beginning allow for some tension. Communication experts tell us that there is always a period of "storming" when a group comes together. Patience and perseverance will be needed during this time as well as throughout. Interreligious dialogue is never easy and often does not bring about results that can be measured.

Early on, it is important to establish some common ground. This can be done by seeing similarities among religious beliefs. A caution here though: what might seem to be similar could be quite different. For instance, "rebirth" in the Hindu tradition is quite different from the Christian belief in "being reborn." Seeing Buddha as divine is quite different from seeing Jesus Christ as divine. Nevertheless, by seriously studying different traditions and comparing them, we can find similarities and parallels that are useful to establishing common ground. For instance, it is helpful to know that the births of Buddha, Krishna, and Jesus are all described as miraculous births. Discovering the universality of religious myths is useful in understanding the commonality among religions. It is useful to compare the power of the Word to connect with God among Sikhs with the parallel notion among Christians, or the Hindus' belief in moksha (liberation) with Christian interpretations. It is interesting to compare the Hindu notion of Atman with the Christian belief in soul, or the Shinto belief in kami with the Native American belief in spirits.

Interreligious dialogue involves a certain "crossing over," a willingness to move into another's world of symbol, story, belief, devotion, and living ritual. We will be walking another's path, trying on their moccasins. It might be compared with going to another country, where the culture, language, and lifestyle are different. Often we have "culture shock" and it takes us a while to ease in and adjust. To be accepted and productive we need to cross over and be open. Otherwise, we remain an outsider and learn little and experience less about this country. It takes courage and flexibility to take advantage of this opportunity to expand and grow. Adapting usually doesn't mean we stay, but we are able to "taste and see" this other land.

When we cross back we somehow are not the same. We have seen our land from a distance, heard other perspectives about it, and now see it differently.

Crossing over into another religion is similar to culture shock. We will experience "shock" crossing over into beliefs we have never heard of. And possibly when we cross back we know that we will never again be the same.

Facing religious differences can challenge our notion of truth. We may think we have truth in a box and that we are absolutely sure of its validity. Then we know that other religions challenge our absolute truths and in a sense make truth relative. Questions arise. Is truth one or many? If truth is one, how can there be so many truths? Are there any absolutes or is everything relative so it doesn't matter what you believe? Just take your pick! Maybe the so-called "post-modernists" are correct when they say that all truth is conditioned by culture, and language is, therefore, relative.

There are no easy answers to these questions. We only turn to some of our sages. As mentioned in the introduction, Gandhi maintained that truth was indeed absolute, but was at the same time a "mystery," a reality that is beyond our comprehension. He spent his entire life searching for truth. As for religious truth, he believed that one must listen to all religions to catch even a glimpse of the truth. In the end, he concluded that the Truth was God, beyond our comprehension. He believed that there were many paths to the same truth. As Monika Hellwig (d. 2005), a great theologian, put it: "There are many paths of salvation, many ways of naming and worshiping the same ultimate, transcendent reality, many languages and rituals by which people search for communion with the divine...."[9]

Approaches to Interreligious Dialogue – Interreligious dialogue can take many forms: one on one, small groups in a class, dialogues among scholars before a class, or guest speakers from various religions. In the following section, you'll be introduced to how digital media and technologies might become integral to your approach to interfaith learning.

The Use of Digital – Today, we live in a digital age with computers, smartphones, and e-readers. Social media is now revolutionizing the way we gain information and communicate. In publishing this book on world religions, we want to make the digital world integral to your learning process.

Twenty-Third Publications will provide a web page (http://store.pastoralplanning.com/woreandcoisi.html) that will lead you in finding many digital multimedia presentations to enhance your learning about other religions and introduce you to many individuals in these religions who are activists in ecology, peace, and women's issues.

To assist you in religious dialogue, university contacts in other countries will be provided so that you can dialogue with people of other faiths. It is also hoped that you will be able to dialogue with individuals and other classes in U.S. and international universities through Cisco and Skype.

SUGGESTED RESOURCES

Banchoff, Thomas, ed. Religious Pluralism. *New York: Oxford University Press, 2008.*

Bryant, M. Darrol and Frank Flinn, eds. Interreligious Dialogue. *New York: Paragon House, 1989.*

Fitzgerald, Michael L. and John Borelli, eds. Interfaith Dialogue. *Maryknoll, NY: Orbis Books, 2006.*

Mays, Rebecca K., ed. Interfaith Dialogue at the Grass Roots. *Philadelphia: Ecumenical Press, 2008.*

Roozen, David A. and Heidi Hadsell, eds. Changing the Way Seminaries Teach. *Hartford , CT: Hartford Seminary, 2009.*

Swidler, Leonard and others, eds. Trialogue. *New London, CT: Twenty-Third Publications, 2007.*

NOTES

[1] *See Leonard Swidler and others, eds.* Death or Dialogue? *(London: SCM Press, 1990)*

[2] *M. Darrol Bryant and Frank Flinn, eds.* Interreligious Dialogue *(New York: Paragon Press, 1989), 6-7.*

[3] *Ibid., 11-16.*

[4] *See Brennan Hill, Paul Knitter, and Wm. Madges,* Faith, Religion and Theology *(Mystic, CT: Twenty-Third Publications, 1997), 198ff.*

[5] *Michael L. Fitzgerald and John Borelli, eds.* Interfaith Dialogue *(Maryknoll, NY: Orbis Books, 2006), 28.*

[6] *Paul Sorrentino,* Religious Pluralism: What Do College Students Think? *(Berlin: VDM Verlag, 2009), 28ff.*

[7] *Rebecca K. Mays, ed.* Interfaith Dialogue at the Grass Roots *(Philadelphia: Ecumenical Press, 2008), 20-24.*

[8] *Fitzgerald and Borelli,* Interfaith Dialogue, *34.*

[9] *Paul Knitter and John Cobb, eds.* Transforming Christianity *(Maryknoll: Orbis, 1999), 51.*